"Despite the apparent novelty of virtual worlds, we forget the power of the text in the construction of imaginary 'real' worlds we can inhabit through reading and travelling. Tourism allows the exploration of fictional landscapes layered upon the real, and MacLeod provides an excellent guide of how these imagined places, peoples and pasts have shaped / continue to shape, the tourist experience."

Mike Robinson, *Professor of Cultural Heritage, Nottingham Trent University, UK.*

Contemporary Geographies of Leisure, Tourism and Mobility

Series Editor: C. Michael Hall, *Professor at the Department of Management, College of Business and Economics, University of Canterbury, Christchurch, New Zealand*

The aim of this series is to explore and communicate the intersections and relationships between leisure, tourism and human mobility within the social sciences.

It will incorporate both traditional and new perspectives on leisure and tourism from contemporary geography, e.g. notions of identity, representation and culture, while also providing for perspectives from cognate areas such as anthropology, cultural studies, gastronomy and food studies, marketing, policy studies and political economy, regional and urban planning, and sociology, within the development of an integrated field of leisure and tourism studies.

Also, increasingly, tourism and leisure are regarded as steps in a continuum of human mobility. Inclusion of mobility in the series offers the prospect to examine the relationship between tourism and migration, the sojourner, educational travel, and second home and retirement travel phenomena.

The series comprises two strands:

Contemporary Geographies of Leisure, Tourism and Mobility aims to address the needs of students and academics, and the titles will be published simultaneously in hardback and paperback.

Routledge Studies in Contemporary Geographies of Leisure, Tourism and Mobility is a forum for innovative new research intended for research students and academics, and the titles will initially be available in hardback only. Titles include:

Tourism and the Spectre of Unlimited Change
Living with Tourism in a Turkish Village Revisited
Hazel Tucker

Literary Fiction Tourism
Understanding the Practice of Fiction-Inspired Travel
Nicola E. MacLeod

For more information about this series, please visit: www.routledge.com/Contemporary-Geographies-of-Leisure-Tourism-and-Mobility/book-series/SE0522

Literary Fiction Tourism

Understanding the Practice of Fiction-Inspired Travel

Nicola E. MacLeod

Routledge
Taylor & Francis Group
LONDON AND NEW YORK

First published 2024
by Routledge
4 Park Square, Milton Park, Abingdon, Oxon OX14 4RN

and by Routledge
605 Third Avenue, New York, NY 10158

Routledge is an imprint of the Taylor & Francis Group, an informa business

British Library Cataloguing-in-Publication Data
A catalogue record for this book is available from the British Library

ISBN: 978-0-367-48581-8 (hbk)
ISBN: 978-1-032-32432-6 (pbk)
ISBN: 978-1-003-04174-0 (ebk)

DOI: 10.4324/9781003041740

Typeset in Times New Roman
by Taylor & Francis Books

To my late father, Duncan Charles MacLeod and my aunt, Elizabeth (Betty) Jane MacLeod.

Both teachers of English, storytellers and my inspiration always.

Contents

Illustrations

Figures

Tables

Boxes

Acknowledgements

I started writing this book as the world closed down in 2020 – a time when there was plenty of literature being read, but very little tourism happening. Since then, it's been a long journey to completion with several unscheduled stop-offs, and I'd like to thank all of those who helped me along the way.

Firstly, many thanks to my editorial team at Routledge: to Emma Travis for her enthusiasm for Literary Fiction Tourism and to Harriet Cunningham, editorial assistant par excellence, for her patient and good-humoured help throughout. Thanks also go to Professor Alastair Morrison for encouraging me to start this project.

To my colleagues at the University of Greenwich, past and present – thank you for your inspiration and friendship. In particular, thanks go to my friends Mel, Elizabeth, Catherine, and James – the ones who got away but are still in touch! To Sam and Ray – thank you for all the years of superlative team-teaching and the fun on our fieldtrips to Malta. To my long-term office mates, Sven, Peter, and Chris, loyal doyens of QA345 – thank you for your company. And to Wenjie and my brilliant, endlessly productive Tourism and Marketing Research Centre colleagues – too many to name, but you know who you are – thank you!

To Lizzie and Jen – bright young things of MA English Literature, University of Edinburgh, 1984–1988, and shining still. Thank you for all the years and for braving a Mallorcan sandstorm to help me photograph the Agatha Christie trail. I'd be lost without you.

To Gerri, my partner in crochet, thank you so much for all the fun, encouragement, and loyal friendship over the years – and for taking me to Haworth and Top Withens.

To Clare LF, your friendship and unwavering interest in all my projects mean so much. Thank you for all our discussions and for your Cannery Row photograph.

To my book group, Julia, Kirsty, Yvonne, Michelle, Clare, and Gerri – thanks for all the chats and for ensuring that I carried on reading stories while researching this book.

To Jonathan and Colin – playwrights and poet – thank you for always being the staunchest of supporters and my very dear friends.

To Kirst and Bowlegs – thank you both for 50 years of friendship – a major achievement!

To my dear family in Scotland: Auntie B, Hughna, Deirdre, Fiona, Lesley, Gillian and all the boys – thank you for always being there through thick and thin, in real life and on WhatsApp. I appreciate you all. I'm also grateful for the memory of my parents, Duncan and June, who still keep me going.

Finally, my gratitude to Johnnie for driving us to all those literary tourism destinations and to Juno for indulging me as I relived my own childhood through our bedtime reading. Thank you to you both.

London

September 2023.

1 The literary fiction tourism landscape

'Frankfurt was delightful. I saw Goethe's house, Schiller's statue … It was very lovely' (Alcott, 2018, p. 368). In a letter home, Amy March, one of the main characters in Volume Two of Louisa May Alcott's *Little Women* (1869) tells her mother of her European literary tourism experiences. In this book, which is part of Alcott's much-loved novel series about a family of young women living in nineteenth century Massachusetts, Amy is accompanying her Aunt Carroll on the American Grand Tour of Europe. This was a touring circuit with which Alcott's readers would have been familiar and which undoubtedly Amy's literary adventures would have further promoted amongst her readership. Orchard House, the home of Alcott and her family, is now a popular visitor attraction in Concord, Massachusetts, a town also famed for its wider literary heritage and associations. The house provided the setting for the imaginary March family home, and visitors to this site enjoy learning about the influential Alcott family through personal artefacts, hearing about their famous literary neighbours, and invoking the fictional characters of Meg, Jo, Beth and Amy March whom they report in reviews that they can sense in every room. Visitors may extend their experience by buying commemorative bags, mugs, and *Little Women* tea in the gift shop and, of course, can purchase copies of Alcott's novels and their film adaptations. There have been seven of these over the years, the most recent being the very well-received 2019 production, directed by Greta Gerwig. The above example clearly illustrates 'the consumption, production, re-production, commodification, transformation, communication, and distribution of literature for tourism purposes' (Robinson and Anderson, 2002b, p.5).

People read literary fiction before, during, and after travel and have long been inspired to visit places because of what they have read. The link between travel and the literary imagination has always been strong with both the fictional outpourings of authors and the interest around their own lives creating an impetus for visitors to tour specific locales. Travel is also a key theme for many writers which creates a synergistic relationship between the activities of reading, writing, and tourism. Literary tourism has the ability to take readers/visitors on a journey to real or imagined places. In these places, authors, and their works, in addition to other literary associations, are the attraction for

DOI: 10.4324/9781003041740-1

visitors. According to Hoppen, Brown and Fyall (2014), 'literary tourism occurs when authors or their literature become so popular that people are drawn to either those locations associated with the author … or those featured within their writings' (p. 37). It is becoming an increasingly popular form of specialist tourism with a diverse range of literary experiences and destinations available for visitors in most regions of the world.

Literary tourism takes a variety of forms, whether the visitor is keen to visit individual sites associated with the life of an author, or the imaginative locations of their fictional worlds. An anthropological interest in the personality behind the creative process may lead us to visit the home which the author inhabited, for example the apartment in Place de Vosges in Paris where Victor Hugo worked on his novel *Les Misérables* (1862). Devotional feelings may inspire us to take a pilgrimage to an author's burial place such as Ernest Hemingway's simple gravestone in Ketchum Cemetery, Idaho, United States of America (USA), where fans still leave coins, flowers, and half-drunk bottles of alcohol in tribute. Or perhaps we might wish to engage with the creative artform of literature itself – this impulse might lead us to the Pennine Moors in West Yorkshire to relive the heightened drama of Emily Brontë's Gothic novel *Wuthering Heights* (1847) in its real-life location. Thus, it is either the content of the art form that inspires the visit or an interest in its creator. Fiction-inspired visitors also enjoy wider book-related themes by attending literary festivals, exploring destinations with a strong literary heritage and well-stocked bookshops, or following literary trails and tours. There is even touristic interest in visiting libraries and iconic literary artefacts such as *The Book of Kells* in Trinity College, Dublin or the *Gutenberg Bible* in the Library of Congress, Washington DC.

England's national tourist board, VisitEngland, stated in a recent report that one in four domestic visitors had visited a United Kingdom (UK)-based literary site in 2016 (VisitEngland, 2017). Overseas visitors also show an interest in these attractions with 5% of all inbound tourists to the UK visiting a literary, music, or media site in 2019 (Visit Britain, 2020). Mintel's 2011 report on international literary tourism predicted that the market will continue to expand with an ageing population boosting visitor numbers (Mintel, 2011). However, despite its recent growth, this form of tourism is not a new phenomenon as shall be seen in Chapter Two. Visitors have been following in the footsteps of their favourite authors as long as there have been writers to trail. However, the study of literary tourism as a form of organised travel practice is relatively new with academic enquiries in the field only being produced since the 1980s. Early classic studies include those of Butler (1986), Herbert (1996), Pocock (1992) and Squire (1994), and a recent review of scholarly articles on literary tourism reveal that the bulk of these publications date from the early 2000s onwards (Çevik, 2020). Most of these articles are based in Anglophone countries and it is clear to see that English Literature and its canon of literary greats still defines much international travel (Robinson and Andersen, 2002b). This is, in part, a legacy of the historical development of this form of touring, but it is also evident that other regions of the world are currently developing their literary tourism and exploiting their associations (see Box 2.1).

Forms of literary tourism

In an attempt to organise this mass of literary-inspired activity, studies have tended to broadly categorise literary tourism resources as either author-centred or fiction-centred (Herbert, 2001; Robinson and Andersen, 2002b; Hoppen *et al.*, 2014). In the author-centred category, places specifically associated with the life of the author are the focus of visitation. Such places comprise writers' homes or graves. Fiction-centred places relate to the imaginative output of the author and include landscapes which inspired literary locations which appear in texts. Within this broad author/fiction distinction, various researchers have attempted to further classify the various literary experiences offered by these locations (see Table 1.1).

The first and most-cited typology of literary place is from Richard Butler's study of 1986 where he proposed four forms of literary tourism, the first being homage to an actual location, for example, a fan of Charles Dickens visiting his house in Doughty Street or his grave in Westminster Abbey, both in London.

Table 1.1 Types of literary tourism

Type of literary tourism	*Features and examples*
From Butler (1986) Aspects of homage to an actual location.	Visits to the authentic homes or burial sites of literary greats e.g., Boris Pasternak's house in Moscow or Robert Louis Stevenson's grave on the island of Samoa.
Places of significance in the work of fiction.	Visits to places which have featured in works of literary fiction e.g., the town of Nikko in Japan, the setting for James Clavell's novel *Shogun* (1975).
Appeal of area because they were appealing to literary and other figures.	Visits to locations much-loved by writers, for example Spain's associations with Ernest Hemingway and Robert Graves.
The literature gains popularity in the sense that the area becomes a tourist destination in its own right.	Visits to regions which become known as literary 'countries' because of their close association with authors and their work e.g., 'Brontë Country' in Yorkshire, 'Steinbeck Country' in California.
From Busby and Klug (2001) Travel writing.	Visits to places that have been showcased in popular travel writing and have thus gained renown e.g., those featured in the work of bestselling author Bill Bryson.
From Busby and Laviolette (2006) Film induced literary tourism.	Visits to literary locations because a tourist has seen a screen adaptation of the literary work e.g., popular film and television adaptations of the works of George Eliot, Charles Dickens, and Jane Austen can induce such visits to film locations.
From Mintel (2011) Literary festivals.	Visits to towns and regions to attend literary festivals, e.g. Jaipur Literature Festival, Rajasthan, India.
Bookshop tourism.	Visits to towns which are renowned for their wealth of book shops e.g., Clunes book town in Victoria, Australia.

Butler later suggests that this homage could take two forms – one might be a literary fan who may wish to see the very desk where a novel was written or they may seek to experience the 'backcloth against which a work was produced to gain new insight into the work and the author' (Butler, 1990, p. 115). The second type of literary tourism proposed by Butler (1986) is to places of significance within the work of fiction itself such as Thomas Hardy's Wessex or John Steinbeck's Cannery Row in Monterey, California (see Fig. 1.1), both of which provide compelling settings for the fictional action in their novels. The third type relates to the draw of an area because it was appealing to literary figures, for example the connections between Spain and the American writer Ernest Hemingway who travelled extensively there from 1936 onwards. The final type is when the area becomes so associated with its literary theme that it becomes a tourist destination in its own right, the most referenced example of this being the creation of the seaside resort Westward Ho! in North Devon, which was developed as a direct result of the popularity of Charles Kingsley's novel of the same name (see Chapter Six). The seaside resort was founded in 1863, nine years after the publication of Kingsley's novel in 1854. The resort is still welcoming visitors and Kingsley's tiny cottage in nearby Clovelly is open to the public (Essex and Brayshay, 2018). Other examples which Butler proposes are the literary 'countries' which emerge around the fame of literary figures and their work within regions. The Lake District and its direct association with the Lake Poets or the Highlands of Scotland, almost singlehandedly brought to the attention of Romantically inclined tourists by Sir Walter Scott, are both clear examples of these 'destinations in their own right.'

Figure 1.1 Cannery Row, Monterey, California. Photograph by C. Le Fort.

Since Butler published his classification in 1986, various authors have proposed additional types of literary tourism to the classification, for example Busby and Klug (2001) added travel writing and film-induced literary tourism, and in their report, Mintel (2011) suggested including literary festivals and bookshop tourism. Travel writing plays an important role in communicating the character of places, and destinations are opened up to a wider public through the lens of travel writers such as Dervla Murphy, Paul Theroux, and Bill Bryson amongst many others. The power of travel writing as 'a channel through which both places as well as people have been re-interpreted and communicated to wider audiences' (Hoppen et al., 2014) is acknowledged here but will not be considered in this book which is focussed on imaginative literary texts. Busby and Klug (2001) and Busby and Laviolette (2006) also propose the inclusion of what they call 'film-induced literary tourism' where visitors have been attracted to a literary location, character, or author after watching a screen adaptation. The various English country houses and estates which have been used as locations for popular television adaptations of the novels of Jane Austen, for example, have experienced extra footfall through this form of inducement. The strong link between screen tourism and literary tourism will be explored in Chapter Nine. Finally, Mintel (2011) proposed that literary-themed events such as festivals should be included in the list of categories as should bookshop tourism which draws visitors to specific destinations because of the quality of the book shop provision and the opportunities for leisurely holiday browsing which they offer. Often these destinations also host literary festivals and other themed events, and these linkages will be examined in detail in Chapter Seven.

In this book, the focus is specifically on literary fiction and the interplay between creators of imaginative writings, their literary settings, and touristic activity. In order to explore these themes, this book will acknowledge the existing studies, but will concentrate on five specific categories of literary fiction tourism which draw on the typologies elaborated above. These categories are: Touristic activity associated with biographical literary places; Imaginative literary places; Socially-constructed literary places; Places of literary repute and literary trails and tours (see Table 1.2). These classifications are explained in more detail in the following section.

Biographical literary places

The first type of literary fiction tourism setting that we will focus on is the biographical place which is considered in detail in Chapter Four. Here we will specifically examine those places and associated artefacts which have an immediate link to the life and death of the author, the most popular of these being writers' former homes and their final resting places. Both spaces offer an immediate connection as the visitor can imaginatively populate the domestic space with their image of their favourite author, look through their eyes at original interiors or views through the window and ponder on their

Table 1.2 Categories of literary fiction tourism used in this book

Categories of literary fiction tourism used in this book	*Examples*
Biographical literary places – places and artefacts closely associated with the author's life (see Chapter Four).	Visits to novelist Katherine Mansfield's birthplace in Wellington, New Zealand.
Imaginative literary places – locations presented within the literary works (see Chapter Five).	Visits to Verona, the Italian setting for Shakespeare's *Romeo and Juliet.*
Socially-constructed literary places – themed sites and destinations deliberately constructed to explore a literary theme (see Chapter Six).	The reconstruction of children's author Beatrix Potter's farm outside Tokyo, Japan.
Places of literary reputation – areas renowned for their literary heritage and associations (see Chapter Seven).	Durban, South Africa as a United Nations Educational, Scientific and Cultural Organisation (UNESCO) City of Literature.
Literary trails and tours – guided and self-guided journeys through the literary landscape (see Chapter Eight).	Agatha Christie Trail in Devon, UK; Don Quixote Route, Castile-La Mancha, Spain.

writing desk, or a lock of their hair. The authenticity of such spaces and objects is key to the power of the biographical place and this is considered in detail in Chapter Three. Graves and monuments are also important biographical sites, offering a tangible place for pilgrims to be in the presence of the remains of authorial genius and offering an opportunity for a more emotional response within the literary tourist. Homes and graves are potent places, but special mention must also be given to birthplaces, favourite holiday locations such as hotels and holiday homes, and workspaces such as libraries and cafés, all of which retain close biographical links to the life of the artists and may also appear in their works. In Chapter Two we consider the 'homes and haunts' genre of literary tourism which extends the writer's domestic sphere into the familiar hinterland of their habitual wanderings.

Imaginative literary places

The biographical site represents the bricks and mortar which houses the literary genius and reflects their life and mode of production, but the imaginary place is a more slippery construct. In Chapter Five, we consider the places which have become the focus of visitation, not because of any direct link to the life of the author but because they have featured in a literary work. These settings are just as powerful as biographical settings as they allow the visitor to inhabit a place that has only hitherto existed in their imagination as they read a work of literary fiction. Compelling characters that dwell in the imaginations of the reader can be brought to life through visiting the fictional settings against which the action takes place, and this can create meaning for visitors. Thus, it is the work of fiction that provides the draw here, rather than details of the life and work of the author. The imaginative setting may be a real place, so meticulously detailed in a work of fiction that the novel might be used as a travel guide by the reader, for example the *Waverley* novels of Sir Walter Scott, set in the Highlands of Scotland, which offer accurate topographical detail for the would-be literary tourist (Aitchison, MacLeod and Shaw, 2002). Alternatively, it may be a composite place that uses real settings, imagined spaces, and renamed locations to create a new literary landscape. These more diffuse landscapes may provide enticing challenges for doughty readers who want to use the books as clues to the physical locations of favourite fictional scenes. Of course, the separation of autobiographical and imaginative spaces is not always as clear cut as this overview might suggest and some authors, such as Lucy Maud Montgomery, author of the popular *Anne* series of books (1908–1921) set in Canada's Price Edward Island, use their childhood or adult homes as inspiration for their literary settings. Watson (2006) suggests that the most poignant and arresting places are those which combine both the autobiographical and the imaginary and such spaces are explored further in Chapter Five.

Socially-constructed literary places

Not all sites of literary fiction tourism relate to the immediate biographical sites of the author's life or the settings for imaginative locales in literary works. Some sites which still attract large numbers of visitors have been deliberately constructed in order to present a literary theme. These sites may exploit and extend an existing literary theme *in situ*, for example a visitor centre on the site of an 'official' literary location which may offer a multi-sensory experience for visitors, involving participation and other forms of interpretation. Conversely, a socially-constructed site may offer de-contextualised experiences based on a literary motif such as those found in theme parks which often use figures from children's literature as the basis of the experience. Such sites are often commercially owned or are developed as part of regional tourism development policies. The presence of these socially-constructed literary sites demonstrates the power of the literary marker as a unifying theme to create identities for places. It also exemplifies the usefulness of such motifs for individual enterprises to attract a range of visitors who may or may not have specific literary motivations. Literary themes also generate the production and consumption of associated merchandise and souvenirs which extend the theme, sometimes rather tenuously, and serve as reminders of the literary encounter. The broad range of literary-inspired products which are sold at literary sites, within the wider literary destination and through online gift retailers, will also be considered in Chapter Nine.

Places of literary reputation

This category of literary tourism refers to visits to locations which are attractive because of their reputation as places which have nurtured writers, and which are renowned for having a prolific literary output. These are settings which are considered to have a high level of literary cultural capital (see Chapter Three) which might come from their reputation for hosting literary events or festivals. They may be renowned for the number of independent bookshops to be found there – these are usually antiquarian or specialist retailers whose array of print books still retain strong appeal despite the rising popularity of e-books, particularly during the global Covid-19 pandemic lockdowns of 2020–21 (Sweney, 2020). This high concentration of book shops gives rise to 'bookshop tourism' (Mintel, 2011) and may earn the location the title of *Book Town*, as designated by the International Organisation of Book Towns (IOBT, 2022). Such book towns often have a high level of rural and heritage appeal and are already on the tourist map. Places with a literary reputation may be valued for their libraries or literary café or pub culture with guided walks being offered to allow visitors to partake of this heady atmosphere. Towns and cities can of course combine all the above and in these cases, they may be eligible for the UNESCO sponsored accolade of City of Literature, the ultimate marker of literary reputation.

Literary tours and trails

This final category does inevitably overlap with the other forms of literary tourism outlined above but it is given its own separate consideration in Chapter Eight because literary tours and trails both offer an opportunity for packaging several different types of literary tourism experience into one coherent product. They also form a linear experience as opposed to the nodal sites which are discussed above and so can represent a journey through the literary landscape. A literary trail is a self-guided and themed walk (or drive) which incorporates several sites or monuments which are associated with an author and their works, and which are open or available to view by the general public. A literary tour is usually a commercial, guided experience (a highly qualified guide being one of the key selling points), and will often offer 'behind the scenes' access to places not generally open to the public. A literary tour or trail will allow a region to direct the tourist gaze (see Chapter Three) and package itself as a literary destination. The physical journey that is created will encourage literary pilgrims to follow in the footsteps of their heroes and, of course, the literary theme offers an attractive and lucrative option for those offering guided walking or motorised tours.

What is literature?

As outlined above, this book will focus primarily on forms of tourism associated with literary fiction rather than travel writing and other forms of tourist-directed text as it is the interplay between imaginative worlds, literary reputation, and the act of touring that is under scrutiny here. It would therefore be useful to explore what is meant by the terms *literature* and *fiction* to examine the power of imaginative writing for the reading public and would-be literary tourist. The use made of the term *literature* in this book, therefore, refers to what is normally called 'creative writing' which includes novels, short stories, poems, and the scripts of plays (Robinson and Andersen, 2002b). The Cambridge Dictionary (2020) defines literature as 'written artistic works, especially those with a high and lasting artistic value' and we shall see that much literary tourism is generated by authors who belong to the literary canon of great and enduring works – although not all. The term *fiction* is linked to literature in that it refers to specific forms of narrative writing (mainly novels) that are based on imaginary events and characters. Works of fiction might take their place within the literary canon, or they may have less lasting appeal and belong to the realms of popular culture.

The idea of a literary canon, a list of the works considered to be the greatest and most influential, was originally an ancient Greek concept when the word kanṓn meant a measuring rod or standard. There have been many attempts at creating and redefining the literary canon, such as Arnold Bennett's *Literary Taste: How to Form It* (1909) and Harold Bloom's *The Western Canon* (1994). Bennett's belief in the significance of literary taste to a life well-lived is summed up in his introduction where we are reminded that 'literature is first and last a means of life, and that enterprise of forming one's literary taste is an enterprise of learning how best to use this means of life' (1937, p.9). These influential compilations have more recently been bolstered by more populist 'bucket lists' of 100 novels to read before you die, published by newspapers and sending anxious readers off to work out how many they still have to contend with. Of course, these lists have predominantly consisted of white European male writers although the canon has recently expanded to include more entries from black writers, women, and from the literatures of Asia, Africa, the Middle East. However, despite the emphasis on the canonical seal of approval, not all fiction-inspired tourism activity is guided by it. As Robinson and Andersen (2002b, p. 28) suggest, 'literary tourism comprises a readership with a will of its own, it does not restrict itself to those works and authors who have somehow won a place on the literary Parnassus.' Thus, for every Shakespeare enthusiast there is also a Catherine Cookson or Dan Brown devotee.

Literary fiction, at its most fundamental level, is the telling of stories which has been a source of pleasure long before any writer put quill to parchment. We enjoy stories because they imitate life as Aristotle believed in his theory of mimesis as an underlying human trait (Woodruff, 1992). We learn from imitation and take great pleasure in artforms which are mimetic and can therefore immerse us in the places, situations, and characters conveyed. The oral tradition of storytelling goes back thousands of years and was used as both a form of entertainment and as a means of passing information down through generations. Since the advent of literacy, we understand that stories as 'literature' refer to a particular patterning of words which we recognise as creative writing. This form of writing is clearly differentiated from other formalised types of written discourse such as newspaper articles or dictionary definitions. This recognisable pattern of vocabulary, metre, rhyme, plot device, and imagery inform us that we are in a fictional world, and it is the skill with which the author creates this pattern that determines our enjoyment of the story created and is also the basis for scholarly appreciation of the work through the discipline of literary criticism. Of course, it is not only the word patterns which denote a work of fiction – we recognise the different formats in which such writings appear, whether they be novels, short stories, poems, or plays, and we can identify the different conventions employed in presenting the text. These literary forms are both conveyors of the story, and, in the format of a book, are also attractive, tactile, and collectible cultural artefacts for personal appreciation. They can also feature as literary tourism exhibits in their own right, for example the earliest surviving copy of *Hamlet* (1603) which is a popular display item at the British Library in London.

Literature and its readers

As a mode of expression, creative writing appears to be much more accessible than other art forms, for example sculpture or opera, as reader and writer share the medium of words, the reader having been generally taught to read from an early age and to recognise the accepted literary greats through the school curriculum. Thus, the reader 'is already in the world of the author, we already share some understanding of the author's linguistic raw material' (Robinson and Andersen, 2002a, p. xiv). The reader, already at home within the art-form of fiction, can immerse themselves in the literary world presented to them. This engagement involves utilising both the imagination and the emotions and the reader willingly puts aside their credulity in their wholehearted acceptance of the fictional world of the novel or the poem (Smith, 2012). Jacobsen (1982) recounts that in his *Preface to Shakespeare* written in 1765, Dr Samuel Johnson believed that we have an emotional response to tragic events in the Bard's plays, not because we believe the eye gouging, poisonings, or drownings to be real but because we empathise with the characters. Similarly, the Romantic poet Coleridge, who coined the term 'willing suspension of disbelief' in his *Biographia Literaria* (1817), believed that this predisposition to accept the fabrication showed 'poetic faith' within the reader – a momentary trust in the fictional world unfolding in the mind (quoted in Jacobsen, 1982, p. 22). It is the skill of the author to create such compelling characters and locations that a reader will want to escape into them both imaginatively, and in the case of literary tourism, in the real-life settings. Literary tourist-readers have visited the locations and met with the characters in their mind – they then travel to literary locations to bring that imaginative experience to life. Similarly, their admiration for the author who has brought these creations into being will also prompt literary-inspired visitors to seek out places and objects associated with the life of the author in an attempt to revive and commune with the creator.

Although we are largely focussed here on the way in which literary fiction affects the leisure reader and potential tourist, it is also useful to consider the academic discipline of literary criticism which is taught in universities across the world. Students learn how to evaluate and interpret works of literature, informed by a series of literary theories which have evolved over time in tandem with developments in other associated disciplines such as sociology, history, and media studies. Alison Booth (2016) informs us that students of literature were encouraged to visit the homes of the authors they were studying, as official academic literary tourists, up until the early twentieth century. Reuben P. Halleck's influential textbook *Halleck's New English Literature* (1900) exhorts its student readers to engage with cultural geography and a literary map of relevant locations was included in the text to facilitate this (cited in Booth, 2016, pp. 22–23). The New Criticism, however, which held sway in British and American universities for much of the twentieth century, attempted to strip the literary work of any authorial context, focussing instead on the purity of the text which should function as a self-contained and self-referential entity. Although recent approaches to literary theory embrace a more pluralistic approach to the interpretation of literature with post-colonial, feminist, social history, and media-related analyses bringing the author back into the fore, there is a lingering discomfort associated with literary ephemera amongst scholars who have been trained to consider interest in the author to be spurious at best and visits to view locks of hair, desks, and graves of writers to be an embarrassing enterprise (Booth, 2016). Indeed, the very existence of literary tourism as the marriage of a highbrow artform with a populist and overtly commercial activity could be considered problematic in some circles. Nicola Watson notes that as literary tourism is 'a practice that tries to make the emotional and virtual realities of reading accountable to the literal, material realities of destination, it is bound to make literary specialists uneasy' (2009, p.5).

Away from the rarefication of university literature departments, fiction is a popular genre within global book sales. At the time of writing, 59 of the top 100 bestselling books on global online retailer Amazon were works of fiction with the next most popular categories being self-help guides and cookery books (Amazon, 2023). Reading novels continues to be a major leisure activity as evidenced by a recent UK government survey which found that one in five Britons (21%) describe themselves as an avid reader and two in five (43%) say they read for pleasure at least once a week. A third of UK readers prefer fiction and their favoured genres are crime and thriller novels which are read by 33% of fiction fans (YouGov, 2020).

For several years, e-readers have offered an alternative to print media for readers. This has improved the affordability of books and access to reading materials. They are also lighter to carry when travelling and are more environmentally responsible. New developments in e-readers are enhancing the reading experience so that it feels more authentic, and the technology now allows e-lending between readers. The global Covid-19 pandemic of 2020–2021 has further encouraged the consumption of electronic reading matter, but statistics show that the traditional printed book still retains its popularity, with Kantar (2021) reporting that 37% of British adults bought a printed book in 2020 and 10% purchased an e-book. Audio books and literary-themed podcasts offer further access to fiction, particularly for readers on the move. The link between travel and reading has always been strong as we shall see in the following chapter. Reading has long been a popular pastime to while away the journeying hours and the time spent waiting at airports. Reading is also a portable leisure activity to undertake whilst on holiday, so much so that book sellers routinely list their top ten holiday or beach reads for this very purpose. Many such travel-orientated novels are about the act of vacationing itself and are aimed at the holiday-making market, for example *Beach Read* by Emily Henry (2020) or *Going Greek* (2021) by Sue Roberts, both of which use the romantic opportunities that travel offers on which to build their plots.

Reading is often seen as a solitary occupation but the rise in the popularity of book clubs and literary societies shows that sharing the literary experience is also valuable. Book clubs have their roots in early Bible reading groups of the seventeenth century with more secular reading clubs appearing from the eighteenth century onwards (McHugh, 2021). These groups have always been particularly associated with women readers because it allowed them to take their place in male-dominated intellectual society and created a space to discuss books with like-minded readers (Burger, 2015). Since the 1990s, celebrity reading groups such as those run by Oprah Winfrey in the USA and Richard and Judy in the UK have wide influence and drive sales (McHugh, 2021). The book club has even spawned self-referential novels about book clubs themselves such as *The Book Club* by Roisin Meaney (2021), *The Book Club* by C.J. Cooper (2021), and *The Jane Austen Book Club* by Karen Joy Fowle (2004). Literary societies are generally focused on a single author and have a more formal constitution. Although they proliferated in the nineteenth century, they still survive and are 'structured around textual correspondence, the circulation of a society newsletter and journal, and the arrangement of a series of annual gatherings' (Reed, 2019, p. 67). They also provide guided tours and expert talks where members can extend the solitary experience of reading with like-minded companions. Within such groups, we also see the production of fan fiction, where published works are expanded upon and reworked by amateur writers who may develop prequels and sequels or change the fate of fictional characters. The online environment has allowed fan fiction to expand significantly, and it now has its own sub-genres, academic scholarship, and a global online archive, FanFiction.

The supply of literary tourism sites

Although there are no statistical data available on the global scale or market for literary tourism, in this book, and in the many other studies in the field, the power of literature to inspire and promote tourism is accepted and tourism to visit sites of literary interest is understood to be a popular activity. As Lee and Weaver (2014, p. 184) state, 'literary tourism has grown in popularity because of demand-side factors of nostalgia, curiosity, reflected place attachment, authenticity seeking, and patriotism and on the supply side because of awareness of the potential of literary resources to stimulate rural peripheries.' That potential to stimulate tourism is also found in urban settings as we shall see throughout the following chapters.

There are very few information sources on the supply of literary-themed sites and on their rates of visitation. Some notable exceptions do exist, however: in terms of writers' home open to the public, it has been estimated by Young (2015) that there are around 50 available in the UK and by Trubek (2011) that there are around 70 in the USA. Wang and Zhang (2017) have estimated that there are 20 writers' homes in China and 13 in Taiwan (see Box 2.1). The richest supply seems to be in France where the Federation of Writers' Houses and Literary Heritage have estimated that there are 393 French literary places with 177 sites being writers' houses available to visitors (Bonniot-Mirloup, 2016). Inevitably, there is a dense cluster of these homes in Paris and similarly we see a concentration of literary associations in London through the Blue Plaque scheme which identifies the homes of illustrious previous tenants. Out of 950 plaques, 215 have been given to writers' homes, showing the concentration of literary associations in the city (see Box 1.1). However only a handful of these operate as visitor attractions and are open to the public.

Box 1.1 Case study: London's Blue Plaque scheme

The London Blue Plaque scheme, which uses distinctive blue ceramic signs to denote buildings of historical interest and those associated with significant figures, is over 150 years old, making it the oldest such project in the world. The MP William Ewart first came up with the idea for commemorative plaques in 1863. The scheme was first supported by the Royal Society of Arts, then the London County Council (later the Greater London Council) and is currently run by English Heritage (Eavis and Spencer, 2018; English Heritage 2021).

The first plaque erected was in honour of the poet Lord Byron in 1867. The literary theme flourished and today 215 out of the 950 plaques in London come under the category of Literature. These mark the buildings where literary figures lived and worked and the plaques proliferate in central London, with 58 in the Royal Borough of Kensington and Chelsea, 57 in the Borough of Westminster, and 37 in the Borough of Camden. Most London boroughs can boast a literary heritage with only eight out of the 32 having no plaques at all. World renowned authors of the literary canon are, of course, featured with plaques commemorating Charles Dickens, Thomas Hardy, George Eliot, Virginia Woolf, and the Romantic poets. We also find writers of popular fiction such as Dorothy Sayers, Georgette Heyer, Gerald Durrell, and Ian Fleming represented. Some of the oldest literary plaques include those marking the Soho home of poet John Dryden who died in

1700 and the Covent Garden plaque for diarist Samuel Pepys who died in 1703. One of the most contemporary literary figures to be commemorated is the novelist Angela Carter who died in 1992 and had a plaque affixed in 2019 to her Clapham home where she lived from 1976.

As a global city, London has played host to numerous international writers: American authors abound, with plaques dedicated to novelists Nathaniel Hawthorne, Herman Melville, Mark Twain and poets Ezra Pound and Sylvia Plath. We also find plaques for the Chinese writer Lao She; the Japanese novelist Natsume Soseki; the Indian poet Rabindranath Tagore; the West Indian writer C.L.R. James; the South African wrier Solomon T. Plaatje; and numerous European writers such as Emile Zola, Ugo Foscolo, Henrich Heine, and George Seferis.

The blue plaques are clear markers of historical and literary interest within the urban landscape and direct the tourist gaze for both dedicated literary tourists and for interested passers-by (Bright *et al.* 2021). Women are under-represented amongst the luminaries celebrated by the scheme as Allison Vale outlines in her a guidebook, *A Woman Lived Here: Alternative Blue Plaques, Remembering London's Remarkable Women* (2018). In this guide, biographies are used to recognise the contribution of those not included in the official listings.

The blue plaques have also been turned into a number of tourist products and experiences, for example there are several guidebooks to the blue plaques, including that produced by English Heritage:

Read (2018) *The Writing on the Wall: 100 Iconic Blue Plaques Commemorating Britain's History*

Rennison (2015) *The London Blue Plaque Guide*

Sumeray (2015) *Track the Plaque: 32 Walks Around London's Commemorative Plaques*

Spencer (2019) *The English Heritage Guide to London's Blue Plaques*

English Heritage also hosts a useful searchable website for the plaques and an app for use on the move. Such is the iconic status of these London markers of historical significance that you can commission your own blue plaque to use at home and blue plaque themed merchandise such as plates, badges, and fridge magnets are available. Hotels such as the Dorchester and Athenaeum have collated blue plaque walks for the benefit of their well-heeled guests.

Information on other types of literary sites is also rather patchy but an honourable exception is the detailed list of literary landmarks in the USA produced by the American Library Association. This list enumerates 187 different landmarks which range from homes, graves, and workspaces to inspirational literary landscapes. New York City and State possess the highest number of sites within the USA with Florida coming a close second (ALA, 2009). Finally, another useful source to help us assess the extent and distribution of literary locations is the fascinating Find a Grave website that informs us that there are 4707 graves or memorials coming under the category of Authors and Writers – these listings span from the grave of metaphysical poet John Donne who was interred in St Paul's Cathedral in London in 1631 to that of bestselling novelist Harold Robbins who was buried in 1997 in Forest Lawn Memorial Park, Los Angeles County (Find a Grave, 2022).

Literary tourism demand

As stated earlier, surveys of international inbound visitors show that 5% of all international tourists to the UK visited a media or music related location which is higher than the number of visitors who attended a live sporting event, or visited a zoo, aquarium, or other wildlife attractions (Visit Britain, 2020). The data also shows the strongest literary regions in the UK to be London, with its associations with Dickens, Shakespeare, and Sherlock Holmes, and the Southwest with its links to author Daphne du Maurier, the *Poldark* series of novels (1945–2002) by Winston Graham, and *Lorna Doone* (1869) by R.D. Blackmore. Visit England's survey of domestic literary tourism in 2016 showed that 55% of Britons 'would be interested in visiting a literary location' and that 24% had done so in the last year (Visit England, 2017).

Another indicator of the popularity of literary tourism (in the UK and Ireland at least) is the wide range of literary guidebooks available. These have a long history as we shall see in the following chapter, but a good supply of contemporary guides to literary tourism are still being published. These range from mainstream guides to literary places such as the *Oxford Guide to Literary Britain and Ireland* (Hahn and Robins, 2008), now in its third edition since 1977, to more specific explorations of imaginary literary spaces. These more focussed works include guides to the literary house, whether that be writers' homes or houses in fiction, guides to specific destinations through the lens of selected authors, and guides to bookshop tourism (see Box 1.2).

Box 1.2 Case study: Contemporary guides to literary fiction tourism

Travel to visit physical literary places has long been augmented by publications such as travel guides and literary maps which provide practical guidance to the visitor in the field. As shall be seen in the following chapter, there is also a history of gazetteers and other texts which promoted armchair travel, underpinning the strong link between literature and place without the need for actual displacement. There is still a healthy market for such publications as can be seen by the range of recent publications on the themes of literary travel and literary landscapes. These tend to fall into the following six categories.

Firstly, there are the simple *literary travel guides* which use maps and directions to guide the visitor along a themed route. Examples of such guides which are currently available include:

- Hamilton and Roberts (2014) *Walking the Literary Landscape: 20 Classic Walks for Booklovers in Northern England*
- ST&G (2020) *Great British Literature Map*
- Connelly (2019) *A Walk with Charles Dickens Along the Thames*

Secondly, we have the category of *guides to literary landscapes* which set out to explore the real places depicted in fictional texts. Using a combination of illustrations, photographs, descriptions, and excerpts, these guides explore the geographical backdrops to favourite novels. These are intended to enhance reading rather than be a practical guide for visitors. Recent examples include:

- Baxter (2019) *Literary Places*
- Sutherland (2018) *Literary Landscapes: Charting the Real-Life Settings of the World's Favourite Fiction*
- Tearle (2017) *Britain by the Book: A Curious Tour of Our Literary Landscape*
- Taggart (2021) *The Book Lover's Bucket List: A Tour of Great British Literature*

The third category is *guides to writers' houses.* These publications underpin the important role played by the writer's home and are explored in detail in Chapter Four. These texts provide an analysis of the inspiration that the domestic settings provide for the writer rather than being visitor guides to specific sites.

- Kennedy and Lee (2020) *Lives of Houses*
- Watson (2020) *The Author's Effects: On the Writer's House Museum*
- Channer (2015) *Writers' Houses: Where Great Books Began*

The fourth category is *guides to fictional houses*: these rather specialist publications explore imaginary homes which play key roles in works of fiction. Settings from *Uncle Tom's Cabin* to *Cold Comfort Farm* are explored, showing the important role that dwellings play in works of fiction.

- Richardson (2019) *The House of Fiction: From Pemberley to Brideshead, Great British Houses in Literature*
- Hardyment (2019) *Novel Houses: Twenty Famous Fictional Dwellings*

The fifth category is *guides to fictional lands*. In these publications, a gazetteer/atlas approach is taken to the exploration of fictional landscapes.

The action is plotted spatially to explore the relationship between literature and place in the imagination. Examples of such guides are:

- Lewis-Jones (2018) *The Writer's Map: An Atlas of Imaginary Lands*
- Miller (2017) *Literary Wonderlands: A Journey Through the Greatest Fictional Worlds Ever Created*
- DeGraff (2015) *Plotted: A Literary Atlas*

Finally, the sixth category, *book shop guides* demonstrate both the popularity of browsing book shops as a leisure activity and the promotional power of a literary heritage in creating a destination brand. This topic is explored in detail in Chapter Seven and examples of such guides currently available are:

- Johnson (2018) *Book Towns: Forty-Five Paradises of the Printed Word*
- Boland (2020) *Bookshop Tours of Britain*
- Strauss (2020) *Do You Read Me? Bookstores Around the World*

Demand for literary-themed travel is also evident in the commissioning of television programmes which use authors and texts as a framework for the depiction of appealing landscapes, backed by readings from well-known presenters. At the time of writing, the British Broadcasting Corporation (BBC) television series *Write Around the World* is being aired with actor Richard E. Grant exploring the landscapes of Italy, France, and Spain in search of classical and contemporary writers and their work. UK television's Channel Five also recently aired *Agatha Christie's England, The Brontës' Britain*, and a short series entitled *In the Footsteps of Charles Dickens, Thomas Hardy and Jane Austen*, all travelogues exploring the homes, lives, and the environments that inspired these much-loved British writers. The popular format of these television series suggests an appetite for literary travel, at least in its vicarious form, across a wide section of the population.

Of course, looking for concrete data on the scale of literary tourism and its market is always going to be a somewhat futile business as, at one level, the inspiration for so much literary-inspired tourism takes place at the subliminal level or at the very least in unofficial and diffuse environments. Childhood or holiday reading may inspire us to visit the rocky coves of the Dorset coast, the beaches of the South of France or to take an American road trip without us ever tracing our urges back to the novels of Enid Blyton, F. Scott Fitzgerald, or Jack Kerouac. Such fiction-inspired visits do not always take place in demarcated sites or monuments and are not recorded as official visitor data.

Having reviewed the rather scant evidence on literary tourism supply and demand, we turn our attention to the question of the visitors themselves – what are these fiction-inspired tourists, who seek out the many different types of experience and product discussed above, looking for in their travels? In Chapter Two we will examine the relationship between cultural tourism and literary-inspired travel to understand how this form of touring fits into this wider activity which, according to the United Nations World Tourism Organisation, accounts for around 40% of all tourist trips undertaken (UNWTO, 2018). There is a very useful literature on cultural tourists' motivations and behaviours (for example, Du Cros and McKercher, 2015; Smith, 2016; Richards, 2021) and there have also been numerous studies carried out to ascertain the characteristics and motivations of the literary-inspired visitor. The earliest of these include Buzard (1993), Lockwood (1981), and Ousby (1990) who focus on the importance of taste and literary awareness amidst literary-inspired travellers or pilgrims using historical accounts, diaries, and journals as source material.

The concept of the tourist as pilgrim is well established in the academic literature and is explored in detail in Chapter Three. The notion of the literary pilgrimage with the reader/visitor as devoted and well-informed supplicant at the shrine of the beloved writer is also a perennial trope, and empirical studies on this research theme from within the tourism discipline quickly followed the earlier studies above. In the early 1990s, the work of Herbert (1996), Pocock (1992), and Squire (1994) used visitor surveys and interviews with contemporary visitors to interrogate this concept of the pilgrim visitor, and Çevik (2020) has ascertained that, in the years between 1997 and 2016, there have been over 33 individual studies on this topic. However, the concept of the literary tourist as pilgrim has not always been strongly proven, for example Herbert (1996) showed that visitors with high levels of literary awareness were very much in the minority in the French town of Cabourg, known for its strong connections to Marcel Proust. Similarly, Busby and Shetliffe (2013) found that only 11% of visitors to Lord Byron's ancestral home Newstead Abbey in England could be categorised as pilgrims. However, Çevik's review of these studies shows that there is a strong link between motivation to visit and literary awareness, with visitor satisfaction increasing with the visitors' levels of knowledge on the literary themes presented. It can be assumed that for those small numbers of well-versed and educated visitors, the literary trip is a pilgrimage which leaves them fulfilled by an experience which has added to their store of knowledge of an author or allowed them to imaginatively inhabit a literary landscape. More recent research on literary tourists' motivations emphasise the need to interact with a literary place in order to enhance the understanding of the work (Baleiro, Viegas and Faria, 2022) and the link between reading practice and education (Yu and Xu, 2023).

The remainder of tourists to these sites may be considered to be more general visitors (Herbert, 2001; Smith, 2003) who are simply accompanying friends and family members as they pursue their own interests. They might be interested in the art, history, and architecture which will also be on display, or they may have recently watched a film or television adaptation of the featured writer's work and been inspired to seek out authentic locations within which to relive the mediated experience (see Chapter Nine). Others may be looking for a day out in pleasant surroundings with the added attraction of a recognisable literary ambience. Depending on the nature of the literary themes being presented, that ambience may include a high level of nostalgia for visitors who are reliving childhood memories through favourite authors and texts. Sheila Squire's well-known study of Beatrix Potter's home in the still untouched surroundings of England's Lake District found that visitors 'repeatedly used Potter and literary tourism to conceptualise and recapture idealized images of childhood and family life' (Squire, 1994, p. 10). To be able to step back in time into that world of fantasy, created when children become absorbed in the literary realm, is a compelling experience and one that draws visitors to literary sites. Future studies of sites associated with a very different literary Potter will no doubt uncover similar levels of nostalgia.

Conclusion and structure of the book

In this introductory chapter, the concept of literary fiction tourism has been introduced and the various forms in which it is found have been outlined. The specific types of literary tourism that will constitute the focus for this book have then been introduced, namely biographical, imaginative, and socially-constructed literary places, regions and towns with a literary reputation, and trails and tours with literary themes (see Table 1.2). The chapter then goes on to define literary fiction and to explore the consumption of fiction as a leisure activity before ending with an overview of the supply of literary fiction sites and experiences and an indication of the levels of demand for these tourism products. Finally, the motivations of the literary traveller, whether they be pilgrims or more general tourists, are considered.

The remainder of *Literary Fiction Tourism* is structured as follows: in Chapter Two, readers will be introduced to the historical development of literary-inspired tourism, from the earliest recorded examples in the Classical era to the present day. Chapter Three situates literary fiction tourism within the wider contexts of cultural and heritage tourism and explores a range of theoretical perspectives through which this activity can be studied. The book then goes on to explore in depth the literary fiction tourism forms introduced above. In Chapter Four, the biographical aspect of literary fiction tourism will be explored with sites and experiences related to the life of the author examined. In Chapter Five, imaginative literary places which feature within the literary works themselves provide the focus of attention. Socially-constructed attractions which exploit the appeal of literary themes are explored in Chapter Six and the role of literary reputation in creating attractive tourist destinations will be examined in Chapter Seven. Literary trails, tours, and itineraries form the focus of Chapter Eight and, in the penultimate Chapter Nine, literary fiction tourism synergies with film, television, and digital culture will be explored. Final conclusions and a consideration of the future of literary fiction tourism will form the basis of Chapter Ten. Throughout the text, current international case studies will allow readers to understand this most ancient of touristic activity within its contemporary context.

References

Aitchison, C., MacLeod, N.E. and Shaw, S.J. (2002) *Leisure and tourism landscapes: Social and cultural geographies.* Abingdon: Routledge.

ALA (2009) Literary landmarks: American Libraries Association. Available at: https://www.ala.org/united/products_services/literarylandmarks (Accessed: 19 March 2021).

Alcott, L.M. (2018) *Little women*, Volume Two. First published 1869. London: Penguin Books.

Amazon (2023) Best sellers in books. Available at: https://www.amazon.co.uk/gp/bestsellers/books (Accessed: 26 June 2023).

Baleiro, R., Viegas, M. and Faria, D. (2022) 'Contributes to the profile of the Brazilian literary tourist: Experience and motivation', *Anais Brasileiros de Estudos Turísticos*, 12, pp. 1–14. Available at: https://sapientia.ualg.pt/bitstream/10400.1/17865/1/2022.06.18_VersA_o_publicada.pdf (Accessed: 15 July 2023).

Baxter, S. (2019) *Literary places.* London: White Lion Publishing.

Bennett, A. (1937) *Literary taste: How to form it.* First published 1909. London: Jonathan Cape.

Bloom, H. (1994) *The Western canon: The books and school of the ages.* New York: Harcourt.

Boland, L. (2020) *Bookshop tours of Britain*. Oxford: Fairlight Books.

Bonniot-Mirloup, A. (2016) 'Tourism and writers' houses, in between places and literature', *Via. Tourism Review*, 9. Available at: https://doi.org/10.4000/viatourism.808 (Accessed: 21 March 2021).

Booth, A. (2016) *Homes and haunts: Touring writers' shrines and countries.* Oxford: Oxford University Press.

Bright, C.F., Foster, K.N., Joyner, A. and Tanny, O. (2021) 'Heritage tourism, historic roadside markers and "just representation" in Tennessee, USA', *Journal of Sustainable Tourism*, 29 (2–3), pp. 428–447. Available at: https://doi.org/10.1080/09669582.2020.1768264 (Accessed: 21 March 2021).

Burger, P. (2015) 'Women's groups and the rise of the book club', *JSTOR Daily Newsletter*, 12 August. Available at: https://daily.jstor.org/feature-book-club/ (Accessed: 23 July 2021).

Busby, G. and Klug, J. (2001) 'Movie-induced tourism: The challenge of measurement and other issues', *Journal of Vacation Marketing*, 7 (4), pp. 316–332.

Busby, G. and Laviolette, P. (2006) 'Narratives in the net: Fiction and Cornish tourism', in Payton, P. (ed.) *Cornish studies.* Exeter: University of Exeter Press, pp. 142–163.

Busby, G. and Shetliffe, E. (2013) 'Literary tourism in context: Byron and Newstead Abbey', *European Journal of Tourism, Hospitality and Recreation*, 4 (3), pp. 5–45.

Butler, R.W. (1986) 'Literature as an influence in shaping the image of tourist destinations: A review and case study', in Marsh, J. (ed.) *Canadian studies of parks, recreation, and foreign lands.* Peterborough: Trent University, pp. 111–132.

Butler, R.W. (1990) 'The influence of the media in shaping international tourist patterns', *Tourism Recreation Research*, 15 (2), pp. 46–53.

Buzard, J. (1993) *The beaten track: European tourism, literature, and the ways to 'culture', 1800–1918*. Oxford: Oxford University Press.

Cambridge Dictionary (2021) *Cambridge dictionary.* Cambridge: University of Cambridge Press. Available at: https://dictionary.cambridge.org/dictionary/english/literature (Accessed: 20 April 2021).

Çevik, S. (2020) 'Literary tourism as a field of research over the period 1997–2016', *European Journal of Tourism Research*, 24. Available at: https://doi.org/10.54055/ejtr.v24i.409 (Accessed: 25 April 2021).

Channer, N. (2015) *Writers' houses: Where great books began*, Marlborough: Crowood Press.

Connelly, R. (2019) *A Walk with Charles Dickens along the Thames.* London: The Literary Map Company.

DeGraff, A. (2015) *Plotted: A literary atlas.* San Francisco: Pulp.

Du Cros, H. and McKercher, B. (2015) *Cultural tourism.* 2nd edn. Abingdon: Routledge.

Eavis, A. and Spencer, H. (2018) 'Risk and reputation: The London blue plaques scheme', in Pellew, J. and Goldman, L. (eds.) *Dethroning historical reputations: Universities, museums and the commemoration of benefactors.* London: University of London Press, pp. 107–116.

Essex, S.J. and Brayshay, M. (2018) 'Devon tourism: The story of the county's economic leviathan', *Transactions of the Devonshire Association*, 150, pp. 177–222.

English Heritage (2021) London's blue plaques. Available at: https://www.english-heritage.org.uk/visit/blue-plaques/ (Accessed: 3 May 2021).

Find a Grave (2022) World's largest gravesite collection. Available at: https://www.findagrave.com/ (Accessed: 10 May 2021).

Hahn, D. and Robins, N. (2008) *Oxford guide to literary Britain and Ireland.* 3rd edn. Oxford: Oxford University Press.

Hamilton, I and Roberts, D. (2014) *Walking the literary landscape: 20 classic walks for booklovers in Northern England.* Sheffield: Vertebrate.

Hardyment, C. (2019) *Novel houses: Twenty famous fictional dwellings.* Oxford: The Bodleian Library.

Herbert, D.T. (1996) 'Artistic and literary places in France as tourist attractions', *Tourism Management*, 17 (2), pp. 77–85. Available at: https://doi.org/10.1016/0261-5177(95)00110–00117 (Accessed: 21 March 2021).

Herbert, D.T. (2001) 'Literary places, tourism and the heritage experience', *Annals of Tourism Research*, 28 (2), pp. 312–333. Available at: https://doi.org/10.1016/S0160-7383(00)00048–00047 (Accessed: 23 March 2021).

Hoppen, A., Brown, L. and Fyall, A. (2014) 'Literary tourism: Opportunities and challenges for the marketing and branding of destinations?', *Journal of Destination Marketing and Management*, 3 (1), pp. 37–47. Available at: https://doi.org/10.1016/j.jdmm.2013.12.009 (Accessed: 24 March 2021).

IOBT (2022) International Organisation of Book Towns. Available at: http://www.booktown.net/. (Accessed: 12 January 2022).

Jacobsen, M. (1982) 'Looking for literary space: The willing suspension of disbelief re-visited', *Research in the Teaching of English*, 16 (1), pp. 21–38. Available at: https://www.jstor.org/stable/40170876 (Accessed: 29 March 2021).

Johnson, A. (2018) *Book towns: Forty-five paradises of the printed word.* London: Frances Lincoln.

Kantar (2021) Are people still reading physical books? Available at: https://www.kantar.com/uki/inspiration/sport-leisure/are-people-still-reading-physical-books/ (Accessed: 3 February 2022).

Kennedy, K. and Lee, H. (eds.) (2020) *Lives of houses.* Princeton: Princeton University Press.

Lee, Y. and Weaver, D. (2014) 'The tourism area life cycle in Kim Yujeong literary village, Korea', *Asia Pacific Journal of Tourism Research*, 19 (2), pp. 181–198. Available at: https://doi.org/10.1080/10941665.2012.735681 (Accessed: 20 May 2023).

Lewis-Jones, H. (ed.) (2018) *The writer's map: An atlas of imaginary lands.* London: Thames and Hudson.

Lockwood, A. (1981) *Passionate pilgrims: The American traveler in Great Britain, 1800–1914*. Vancouver: Fairleigh Dickinson University Press.

McHugh, J. (2021) 'How women invented book clubs, changing reading and their lives', *The Independent*, 29 March. Available at: https://www.independent.co.uk/life-style/women-book-clubs-reading-america-b1823819.html (Accessed: 15 July 2021).

Miller, L. (2017) *Literary wonderlands: A journey through the greatest fictional worlds ever created.* London: Thames and Hudson.

Mintel (2011) Literary tourism–international–September. London: Mintel International Group.

Ousby, I. (1990) *The Englishman's England: Taste, travel and the rise of tourism.* Cambridge: Cambridge University Press.

Pocock, D. (1992) 'Catherine Cookson country: Tourist expectation and experience', *Geography*, 77 (3), pp. 236–243. Available at: https://www.jstor.org/stable/40572194 (Accessed: 2 February 2020).

Read, M. (2018) *The writing on the wall: 100 ionic blue plaques commemorating Britain's history.* London: Unicorn Press.

Reed, A. (2019) 'Reading minor characters: An English literary society and its culture of investigation', *PMLA*, 134 (1), pp. 66–80. Available at: https://doi.org/10.1632/pmla.2019.134.1.66 (Accessed: 20 May 2020).

Rennison, N. (2015) *The London blue plaque guide.* 4th edn. Cheltenham: The History Press.

Richards, G. (2021) *Rethinking cultural tourism.* Cheltenham: Edward Elgar Publishing.

Richardson, P. (2019) *The house of fiction: From Pemberley to Brideshead, Great British houses in literature.* London: Unbound Press.

Robinson, M. and Andersen, H.C. (2002a) 'Introduction', in Robinson, M. and Andersen, H.C. (eds.) *Literature and tourism: Reading and writing tourism texts.* London: Continuum, pp. xii–xix.

Robinson, M. and Andersen, H.C. (2002b) 'Reading between the lines: Literature and the creation of touristic spaces', in Robinson, M. and Andersen, H.C. (eds.) *Literature and tourism: Reading and writing tourism texts.* London: Continuum, pp. 1–38.

Smith, K.A. (2003) 'Literary enthusiasts as visitors and volunteers', *International Journal of Tourism Research*, 5 (2), pp. 83–95. Available at: https://doi.org/10.1002/jtr.419 (Accessed: 3 October 2021).

Smith, M.K. (2016) *Issues in cultural tourism studies.* 3rd edn. London: Routledge.

Smith, Y. (2012) '*Literary tourism as a developing genre: South Africa's potential*', PhD thesis, University of Pretoria, Pretoria. Available at: http://hdl.handle.net/2263/29043 (Accessed: 20 May 2020).

Spencer, H. (2019) *The English Heritage guide to London's blue plaques.* Tewkesbury: September Publishing.

Squire, S.J. (1994) 'The cultural values of literary tourism', *Annals of Tourism Research*, 21 (1), pp. 103–120. Available at: https://doi.org/10.1016/0160-7383(94)90007-90008 (Accessed: 1 July 21).

ST&G (2020) *Great British literature map.* Devon: Marvellous Maps.

Strauss, M.J. (ed.) (2021) *Do you read me? Bookstores around the world.* London: Gestalten.

Sumeray, D. (2015) *Track the plaque: 32 walks around London's commemorative plaques.* Nottingham: DB Publishing.

Sutherland, J. (ed.) (2018) *Literary landscapes: Charting the real-life settings of the world's favourite fiction.* London: Modern Books.

Sweney, M. (2020) 'Pandemic drives ebook and audiobook sales by UK publishers to all-time high', *The Guardian*, 14 November. Available at: https://www.theguardian.com/books/2020/nov/14/pandemic-drives-ebook-and-audiobook-sales-by-uk-publishers-to-all-time-high-covid (Accessed: 4 June 2021).

Taggart, C. (2021) *The book lover's bucket list: A tour of Great British literature.* London: British Library Publishing.

Tearle, O. (2017) *Britain by the book: A curious tour of our literary landscape.* London: John Murray.

Trubek, A. (2011) *A skeptic's guide to writers' houses.* Philadelphia: University of Pennsylvania Press.

UNWTO (2018) *Tourism and culture synergies.* Madrid: UNWTO. Available at: https://doi.org/10.18111/9789284418978. (Accessed: 12 June 2020).

Vale, A. (2018) *A woman lived here: Alternative blue plaques, remembering London's remarkable women.* London: Robinson.

Visit England (2017) Year of literary heroes – Survey results summary. Available at: https://ftnnews.com/images/documents/2017/year_of_literary_heroes_-_survey_results_summary.pdf (Accessed: 3 August 2021).

Visit Britain (2020) Activities in Britain's nations and regions, *Foresight* – issue 165. Available at: https://www.visitbritain.org/activities-undertaken-britain (Accessed: 3 August 2021).

Wang, H.J. and Zhang, D. (2017) 'Comparing literary tourism in mainland China and Taiwan: The Lu Xun native place and the Lin Yutang house', *Tourism Management*, 59, pp. 234–253. Available at: https://doi.org/10.1016/j.tourman.2016.08.008 (Accessed: 1 July 21).

Watson, N.J. (2006) *The literary tourist: Readers and places in Romantic and Victorian Britain.* Basingstoke: Palgrave MacMillan.

Watson, N.J. (2009) 'Introduction' in Watson, N.J. (ed.) *Literary tourism and nineteenth century culture.* Basingstoke: Palgrave, pp. 1–12.

Watson, N.J. (2020) *The author's effects: On writers' house museums.* Oxford: University of Oxford Press.

Woodruff, P. (1992) 'Aristotle on mimesis' in Oksenberg Rorty, A. (ed.) *Essays on Aristotle's poetics.* Princeton: University of Princeton Press, pp. 73–95.

YouGov (2020) YouGov profiles: Book reading. Available at: https://yougov.co.uk/topics/arts/articles-reports/2020/03/05/world-book-day-britons-reading-habits (Accessed: 12 July 21).

Young, L. (2015) 'Literature, museums, and national identity: Or, why are there so many writers' house museums in Britain?', *Museum History Journal*, 8 (2), pp. 229–246. Available at: https://doi.org/10.1179/1936981615Z.00000000052 (Accessed: 10 July 21).

Yu, X. and Xu, H. (2023) 'Text-induced travels: A self-determination perspective', *Annals of Tourism Research*, 101, pp. 1–12. Available at: https://doi.org/10.1016/j.annals.2023.103599 (Accessed: 15 July 2023).

2 The historic evolution of literary fiction tourism

Writers, the written word, and the literary imagination have always been an inspiration for as long as humans have left their homes for the purposes of travel. In this chapter, the historical evolution of literary-inspired travel is explored, and the changing perspectives and emphases are discussed along the route. In particular, the shift from literary-inspired travel as the domain of the educated elite to its more democratic form will be examined. Robinson and Andersen note that:

> the spaces and places of literature have long been noted as sites to visit and engage with … A persistent and in the main accurate view is that spatial expressions of the lives and works of creative writers have largely held appeal to the traveller rather than the tourist.
>
> (2002b, p. 12)

However, as we move through the eighteenth and nineteenth centuries, we shall see that 'this period saw the practice of visiting places associated with particular books in order to savour text, place and their interrelations grow into a commercially significant phenomenon' (Watson, 2006, p. 1).

The chapter begins with the earliest recorded examples of literary tourism in the Classical and Medieval periods before moving on to an examination of the aristocratic European Grand Tour where engagement with the classics and contact with famed authors formed a mainstay for wealthy and educated literary tourists. The increasing democratisation of tourism through the nineteenth and twentieth centuries is outlined as literary tours became more readily available to a wider franchise who travelled to gaze upon an increasingly diverse range of literary sites and experiences. The lure of authors and literary associations is now recognised by regional and national tourist boards who use these themes to develop attractive identities and encourage tourism. Thus, we will see that the story of literary tourism has been one of a shift from individual travel, inspired by the desire for knowledge and communion with an author, towards the deliberate creation of literary-themed attractions and experiences for a wider range of visitors who may be incorporating these activities within a wider holiday experience.

DOI: 10.4324/9781003041740-2

The movement is therefore from the individual to the mass experience and towards an increasingly commodified offering. Santesso (2004, p. 379) aligns this shift to developments within the history of tourism itself which demonstrate 'an older discourse-centred idea of travel – what we might call interrogatory tourism – and a newer emphasis on looking at managed historical sites and monuments – what would become modern tourism.' The evolution of literary tourism must therefore be understood in terms of wider developments within the twin activities of reading and travel and the chapter briefly outlines some key stages in the history of commercial tourism and in the widening of literacy and the cultural practice of reading.

Classical and Medieval literary tourism

Although the birth of literary tourism has been pinpointed to the Medieval period (Herbert, 2001; Mansfield, 2015) or the European Grand Tour (Santesso, 2004; Schaff, 2009), the earliest forms of literary-inspired travel can be traced much further back to Classical civilizations (Robinson and Andersen, 2002a). One of the earliest stories in existence, the *Tale of the Shipwrecked Sailor*, is a Middle Kingdom (1985–1773 BC) account of an Ancient Egyptian voyage (Abbas, 2013) which establishes the archetypal desert island trope a thousand years before Homer's Odyssey gave its name to all epic voyages of discovery (Hulme and Youngs, 2002). Travel as a subject of writing, and subsequently as a consequence of it, was therefore established in ancient times when literacy was highly prized, and writers were visited out of admiration for their lives and their work. Hendrix (2009) notes that visitors would be affected by being in the presence of writers or in places associated with the lives of deceased authors, and that most of the written evidence of Classical cultural tourism consists of travellers' accounts of visiting biographical sites. These included Cicero's house in Rome and Virgil's grave near Naples which was reputedly an attraction from the time of his death in 19 BC. St Paul was allegedly a noteworthy visitor and was found in tears by the tomb (Watson, 2006, p. 32; Booth, 2016). Hendrix (2009) asserts that at this stage in the development of literary tourism, the focus was not yet on the imaginative locations of the fictional work but instead attention was fixed on the figure of the artist and places associated with their life. However, Quinteiro and Baleiro (2018) identify some textually-inspired tourism during the Roman Empire with visitors travelling to the site of Ancient Troy to relive Homer's epic poem *The Iliad* with help from a tour guide. As little remained of the physical landscape setting, literary references were utilised to help visitors locate the sites of key events (Feifer, 1985).

In the Medieval period, travel was dominated by the religious pilgrimage and written guides exist from as early as the 1100s, for example the 1140 guide to the Camino de Santiago, the *Codex Calixtinus* and the fourteenth century Camino pilgrim guide both of which include information on safe drinking water, ferries, and descriptions of sites to be seen along the route.

The latter guidebook still resides in the British Library in London and can be inspected by contemporary pilgrims. The pilgrimage has also left a literary legacy as it provided the plot device for Geoffrey Chaucer's highly influential *Canterbury Tales* (1387–1400) and John Bunyan's equally significant later work, *The Pilgrim's Progress* (1678). Pilgrims were of course motivated by devotion, but the literary dimension of this form of touring was also compelling, with early admirers of the *Tales* following in the footsteps of Chaucer's colourful and sometimes bawdy pilgrims by replicating their journey from Southwark in London to Canterbury in Kent. Watson (2006) notes that Chaucer's tomb at Westminster Abbey was already a draw for visitors by the 1550s, prefiguring the attraction of the Abbey's Poets' Corner which was to come in later centuries. Mansfield (2015) describes Medieval travel to witness the Mystery plays staged in European towns such as York, London, and Paris as another early form of literary-inspired travel with religious underpinnings. Interest in these texts is still evident – present-day literary pilgrims can follow the John Bunyan Trail in Bedfordshire, United Kingdom (UK) and up until recently could have visited the Canterbury Tales visitor attraction. The city of York also still periodically stages the Mystery Plays to enthusiastic audiences (York Festival Trust, 2023).

In fourteenth century Italy, the Renaissance writer and poet Francesco Petrarch, a keen literary tourist who visited the sites of Virgil's Aeneid in Baia in 1343, became himself a subject of literary pilgrimage after his death in 1375 (Hendrix, 2009). His learned works in Latin earned him respect and infrequent visits to his tomb near Padua, but it was his *Canzoniere*, the emotional lyrics on his infatuation for the mysterious Laura, written in the more immediate vernacular Italian, rather than Latin, which inflamed a fully-fledged literary tourism industry in the 1520s (Quinteiro and Baleiro, 2018). Hendrix (2009, p. 15) explains that it was 'the allegedly autobiographical nature of Petrarch's love poetry that triggered an essentially new kind of readerly interest dominated by curiosity' and this interest led visitors to Fontaine-de-Vaucluse in France where Petrarch is thought to have written his poems, and where he set his love story. Further touristic trappings followed – a literary location map was produced in 1525 and collections of artefacts, such as fragments of Laura's putative chamber pot, were assembled and toured. These examples demonstrate that in this period, the biographical literary site had begun to give way to a wider appreciation of the imaginative settings of literary works, especially those imbued with romance and emotions into which visitors hoped to tap. This impetus would continue to expand, as the following discussion of subsequent periods reveals.

The Grand Tour

The Grand Tour enjoys a special place within any historical account of tourism and, in particular, of literary-inspired travel. Aside from the great Medieval pilgrimages across Europe and the Holy Land, the Grand Tour was the first form of institutionalised travel with its own appurtenances, routes, and itineraries available to potential Tourers (Black, 2010; Towner, 2002). The Grand Tour was predominantly an institution for young, aristocratic men travelling with tutors on extended trips across Europe to complete their education. However, the routes were also followed by prosperous art collectors and noble families demonstrating their wealth through travel which afforded them the opportunity to socialise with their Continental peers. The tradition flourished from the 1600s until the nineteenth century when the widening railway network allowed the middle classes to take shorter, recreational trips to the Continent causing the exclusivity of the Grand Tour to wane. The popularity of the Tour meant that there were many published accounts of individual travels of varying degrees of quality and these travelogues were chief amongst the Grand Tour paraphernalia consulted by would-be travellers. There were useful and well received guides such as Addison's *Remarks on Several Parts of Italy* (1705) and Thomas Nugent's multi-volume *The Grand Tour* (1749), but the market also became flooded with mundane travelogues which became a focus for satirical comment. Towner (2002) recounts an exchange between that well-travelled pair, James Boswell, and Dr Samuel Johnson in 1778, when Boswell's proposal to publish an account of his travels was strenuously discouraged by his friend who queried what he could possibly add to the accumulated knowledge of such a well-trodden part of the world. It would seem that a Grand Tour travelogue was already a cliché by the late 1770s.

Travellers to the Continent were also keen to seek out the company of writers themselves – visiting the German author Goethe was a popular activity and he entertained visitors at his parents' home in Frankfurt. Similarly, Voltaire was at home to travellers in his Villa Les Delices in Geneva and received so many visitors there that he referred to himself as 'the Innkeeper of Europe' (Hendrix 2009, p. 20). Consorting with the *literati* of Europe seemed to have been one of the key benefits of the Tour for some travellers: Black (2010) gives us an extract from a journal of one Mr Petty who, on traveling through Leipzig with his wife, comments that: 'I have made some valuable literary acquaintances and found the society of men of letters very advantageous' (p. 8).

The Tour was therefore always heavily associated with writers and the written word in the form of what we would now call travel writing, but more literary motivations were also significant and Smith (2013, p. 140) suggests that 'reading was one of the most important facets of travel preparations' for Grand Tourists. Tutors and their young noble charges would have been well-versed in the Classics as this was a staple of aristocratic education and an Oxford or Cambridge degree. Therefore, the Europe that they sought was the continent of Classical antiquity and they went in search of the Italy of Horace, Catullus, and Virgil with copies of their works as guides. Towner suggests that up until the mid-eighteenth century, the Grand Tour had a wholly Classical ambience from the routes and sites visited, to the sentiments expressed in describing these in letters, diaries, and published accounts (Towner, 1985). Thus, what Towner refers to as the Classical Grand Tour was carried out by tourists seeking discourse in the form of acquiring languages and meeting and conversing with eminent figures

(Adler, 1989). However, this Classical sensibility gave way in the eighteenth century to the Romantic Grand Tour which was more philosophically and intellectually inspired by the Romantic Movement and importantly, was more directed at visual appropriation of toured places – in other words, sightseeing.

The Romantic Movement heralded a shift in values that abandoned the rational and instead gave supremacy to feelings and the power of nature. Fuelled by a backlash against the Enlightenment and the Industrial Revolution and spurred on by the emancipation of the common man and woman represented by the French Revolution, Romanticism prioritised emotions and the imagination and saw nature as the true guiding-force for humankind (Aitchison, MacLeod and Shaw, 2002). Artistically, the Romantic Movement was represented by the work of poets and writers such as Wordsworth, Byron, Shelley, Keats, Hugo, Goethe, Schiller, and Rousseau. Visual artists of the Romantic Movement included Friedrich, Delacroix, Turner, and Constable, who also chose to represent in their work the realities and power of nature rather than allegorical or Biblical scenes. In wider aesthetic terms, the Romantic Movement was also associated with the Picturesque and Sublime appreciation of landscapes. The Picturesque was a way of seeing landscape as if it were a picture, framed and ready to be consumed. This aesthetic tended towards ruins and decay whilst the Sublime gravitated towards the majestic and awe inspiring in nature, generating intense feelings within the viewer. Punter (1994, p. 223) explains the difference between these terms thus: 'The Picturesque frames roughness and variety: the Sublime has to do with being overwhelmed, surprised, being taken out of one's frame by a scene.' Wild and mountainous landscapes once feared or considered ugly in contrast to domesticated countryside now became prized both as a subject for art and for touring. These were landscapes that elicited an emotional response and underlined the healing powers of nature in an increasingly industrialising Europe.

Tourism and the Grand Tour were inevitably influenced by these trends and soon the heightened emotions of works by authors such as Dante and Petrarch and their thrilling locations were setting the tone and the route rather than the journeys described in Classical texts. By far the most popular text of the early Romantic period was Jean-Jacques Rousseau's *Julie, or The New Héloïse* published in 1761. This emotionally charged epistolary novel explores an illicit affair based on the Medieval love story of Héloïse and Abelard and set in Vevey in Lake Geneva, Switzerland, an area already popular with Grand Tourists and with visitors in search of the Sublime in the Alpine landscape. The book was one of the best-selling novels of its time and the response to its romance and tragedy was immediate – Rousseau identified himself as 'editor' of the letters that comprise the novel which encouraged readers and later, visitors to believe that this was a true story. This putative veracity, and the presence of real landscape settings described in the novel, led to tourists arriving in ever-increasing numbers to explore the locations, searching for the best vantage points around the lake to give them the appropriate emotional response. Nicola Watson suggests that Rousseau's decision to set his novel in realistic settings was 'revolutionary' and visitors 'could and did, plot the novel on a map of the eastern end of the lake' (Watson, 2006, p. 140). Thus, the novel acted as a sentimental guide to the area with visitors reading out relevant sections of texts at appropriate spots, a habit that was soon to become subjected to commercial exploitation as we shall see in the following section.

The nineteenth century tour

The Grand Tour continued to be influential into the next century but slowly evolved in terms of the length of the journey and the social profile of those touring. The Romantic Movement still held sway and the much-travelled Lord Byron became a celebrity of the Tour itself, so much so that 'his Childe Harolde's Pilgrimage or the Prisoner of Chillon could be found clutched in the hands of a new generation of visitors just as Horace and Livy had been a hundred years before' (Towner, 2002, p. 233). These works would take fans on Romantic journeys across Europe in the footsteps of the touring Harolde or to Castle Chillon in Switzerland, comparing Byron's poetic descriptions with *in-situ* sightseeing. This popular practice of reading on the spot led to the production of pocket-sized tourist editions of favourite texts – the works of Byron and Walter Scott were soon available in handy travel sizes to facilitate the literary tourist and souvenir editions were also produced as lasting mementoes of the visit (Goodman, 2019). The desire to read a literary passage in an appropriate setting was also exploited by the production of travel guides. From 1836, John Murray's London-based publishing house produced travel handbooks that were the forerunners of modern travel guides. His *Handbook for Travellers on the Continent* (1836) set out to combine both practical and literary associations connected with those destinations. Aimed at the emerging middle-class traveller, the *Guides* modelled cultured travel for visitors who 'would be more familiar with literary presentations of foreign places than with the yet unknown matter-of-fact description that would become typical of the guidebook as a genre' (Schaff, 2009, pp. 108–109). The Murray guides also provided a portable and efficient means by which visitors would always have some literary inspiration to accompany their European travels.

The UK was also attracting Romantically inspired literary visitors in this period, most notably the Scottish Highlands, which had become transformed through the pen of Sir Walter Scott, and the Lake District which had been written into being by William Wordsworth. These rather remote regions of Britain had become more accessible by the 1800s with the development of the steamship service, the railway network, and the concomitant Thomas Cook's tours. A Cook's Tour package greatly facilitated travel by navigating the tricky business of train bookings and connections and provided a guide to accompany tourists. Thomas Cook, a committed member of the Temperance Movement, originally started arranging excursions to provide a wholesome alternative to the gin palaces of the industrial cities and was soon organising tours around the UK, Europe, Africa, and eventually across the globe with his world tour lasting 222 days. Promotional material for his 'Tartan Tours of Scotland' made many references to the work of Walter Scott and clients were expected to be well-versed in his works (Gold and Gold, 1995, p. 103). According to Buzard, (1992, pp. 52–53) these two-week long tours came to form the biggest part of Cook's business and he claimed to have escorted 40,000 visitors to Scotland by 1861 (Gold and Gold, 1995, p. 103). Whether they came on an organised excursion or by independent means, the Trossachs region in the West Highlands, already a popular scenic destination due to its Picturesque credentials, was their goal. This area, including beautiful Loch Katrine, became a tourist honeypot as it formed the setting for Scott's long poem *The Lady of the Lake* (1810), one of his most popular texts and certainly the most commercially successful (Nestor, 2020). Visitors brought their travel-sized copies of the poem to read in appropriate vantage points around the loch just as Rousseau's fans had done on Lake Geneva in the previous century. Sight-seeing wagon tours were organised with specially trained drivers who were able to recite Scott's verses to passengers. It was said that American tourists were so great in number that the star-spangled banner was flown at the local hotel (Aitchison et al., 2002, p. 86). A small, wooded island situated in the southeast end of Loch Katrine was named 'Ellen's Isle' after the heroine of the poem and this topographical feature still bears this name on the Ordnance Survey map. Loch Katrine is now owned by Scottish Water as the main reservoir that serves the city of Glasgow and so we see practical functionality meeting imaginative geography in this historic literary landscape.

Sir Walter Scott's Baronial mansion, Abbotsford, situated in the Scottish Borders and remodelled in appropriately Gothic and Romantic style, was also a draw for visitors and was the first writer's home to be opened to the public when they were first welcomed in 1833. Further south in the Lake District, in England's northwest corner, the poet William Wordsworth was also providing an attraction, both in terms of his literary settings and his own persona. Wordsworth was born in the Lake District and used this dramatic setting for his poetry which celebrated the redemptive power of nature. His home, Rydal Mount, where he lived with his family from 1813 until he died in 1850, became a shrine even during his lifetime, with visitors keen to spy him and even steal cuttings from his terraced garden which became a famous spot from which to share the same vantage point as the revered poet (Donaldson, Gregory and Murrieta-Flores, 2015). He published his famous *Guide Through the District of the Lakes* in 1810 – a handbook for people of taste and with a feeling for landscape. This region of northern England with it dizzying peaks and large bodies of water was already a favoured spot for Romantically oriented visitors and, as Westaway (2013) suggests, provided a more cost-effective version of the Alpine experience for the less well-off middle-class traveller.

The homes, birthplaces, and burial sites of other famed authors were also the subject of touristic interest. Shakespeare's Stratford-upon-Avon (arguably the first site of literary pilgrimage in England) had been attracting visitors since David Garrick's Shakespeare Jubilee of 1769 and the subsequent establishment of the Shakespeare Birthplace Trust in 1847. In Scotland's Ayrshire, the Robert Burns industry had been functioning since not long after the poet's death in 1794 with his birthplace and other associations, including his surviving family, providing points of interest for the literary traveller (Gold and Gold, 1995; Bhandari, 2008). The compelling lives and work of the famously talented Brontë family in Haworth, Yorkshire were also bringing in the curious to view the home environs and bleak surrounding moors during Charlotte's lifetime – something the last surviving sibling found rather ridiculous (Barnard, 2002). After her death, visitors were further encouraged in no small measure by her vivid biography written by fellow author Elizabeth Gaskell in 1857.

American visitors in Europe travelled for shorter periods on their version of the route but this Tour still had a distinctly literary overtone with visitors seeking out locations associated with their European reading habits. Perhaps inevitably, the UK was a major draw for American literary tourists, as exploring sites associated with British authors and their works was a popular way to relate to the Old Country and share a literary heritage as many would have grown up reading English novels (Westover, 2009). Foster (2009) suggests that written accounts of these visits show that they sought to 'pay due homage to an old world aesthetic while at the same time upholding their own … sense of national independence' (p. 183). Later, that independence would manifest itself in a developing interest in home-grown literary tourism sites with Thoreau's Waldon Pond and the nearby literary town of Concord in Massachusetts becoming a pilgrimage site for thousands of American and British travellers (Buell, 1989) and the Florida home of Harriet Beecher Stowe attracting many visitors on specially commissioned steamboats (Roberts, 2009).

The American on tour in Europe is firmly embedded in popular culture with European travel being used as a literary device in many novels, for example in Henry James's *Daisy Miller* (1878) or *Portrait of a Lady* (1881) and in semi fictional travelogues such as Mark Twain's *The Innocents Abroad* (1869) and *A Tramp Abroad* (1880). As we saw in Chapter One, the American author Louisa May Alcott travelled to Europe in 1865–1866 at the age of 32 as a companion to an invalid friend. During this trip she visited the popular Grand Tour destinations of Germany, Switzerland, France, and England. Her experiences were used in her immensely popular novel *Little Women* (1869) where her character, the artistic Amy March, finds inspiration and romance whilst touring in Europe with her aunt. Similarly, another American author Susan Coolidge used her own European experiences in her novel *What Katy Did Next* (1886). In this classic, much-loved tale, younger readers are treated to Katy Carr's literary pilgrimage through what she refers to as 'Storybook England' where her impressive grasp of the English classics enlivens her tours around London and the South of England, although sadly heavy rain prevents their Scottish jaunt. Katy finally (and inevitably) finds love in Italy, that emotionally charged destination which also brought romance to English literary heroines such as E.M. Forster's Lilia Herriton in *Where Angels*

Fear to Tread (1905) and Lucy Honeychurch in *Room with a View* (1910). These novels are clearly both products of the nineteenth century literary Grand Tour and have also played an important role in replicating its features for successive generations of readers to enjoy on the page and in their own European travels.

Homes and haunts

A further influential development in the historical evolution of literary tourism was the creation of the 'homes and haunts' genre of travel guide which still resonates with contemporary literary travel. William Howitt published his *Homes and Haunts of the Most Eminent British Poets* in London in 1847 and this text became a hit with the reading and travelling public. The guide is a series of essays based on biographical research and visits that Howitt and his wife Mary made to various living authors throughout the 1840s. Volume One includes Chaucer, Shakespeare, Milton, Keats, and Byron and descriptions of their homes, environs, monuments, and graves are included along with excerpts from their work. In Volume Two we hear first-hand accounts of the Howitt's visits to Wordsworth's home at Rydal Mount in the Lake District where the guide refers to the poet as

> being in the very centre of a region swarming with tourists and hunters of the picturesque, and in the very highway of their route is regularly beset by them. Day after day brings up whole troops of them from every quarter of these kingdoms, and no few from America.
>
> (Howitt, 1875, p. 331)

Doubtless the *Homes and Haunts* guides would have subsequently added to the swarm. As a rather eclectic mix of biography, travelogue, and memoir, the volumes' intention was to guide the visitor around the houses and locales that sheltered and inspired famous writers and the text was used both as a guide in the field and as vicarious armchair travel. An American counterpart, Putnam's *Homes of American Authors* was published in 1853 (Hazard, 2016) and the homes and haunts concept became so popular that it gave its name to a genre that is still used and recognised today. Indeed, in the Library of Congress Classification system which dates from 1897, 'homes and haunts' continues to be used as a classifying category (Young, 2019).

Given the specific factors underpinning the development of literary tourism, the above discussion has focussed on studies of European and American travellers' accounts of their activities in these regions. However, there are, of course, examples of literary tourism development across the world, one of which can be seen in Box 2.1, where the evolution of literary-inspired travel in China is outlined. An example of Antipodean fiction-based tourism can be found in the case of English-born author Samuel Butler who wrote his utopian novel *Erewhon* (1872) whilst working on his sheep farm in the Upper

Rangitata Valley in the high country of Canterbury on New Zealand's South Island. The satirical novel was both mythical and topographically accurate and the tale of a Utopian society nestled in the mountains became a best seller. Buchmann (2006) comments that within weeks of its publication *Erewhon* was attracting both domestic and English visitors to see the mountainous regions described in the novel and to visit and collect souvenirs from the author's home which he had already vacated by this time. Butler's cottage is long gone but a plaque has reminded visitors of its location since 1950 and the spot is now an official New Zealand Heritage Site. A mountain range and peak have been named after Butler; thus, the utopian tale and its author have been incorporated into the landscape and the history of the region.

Vogel (2002) reminds us that the Grand Tour reached out beyond the confines of European capital cities in her exploration of the Gothic and sublime in African tourism. The novelist Oliver Goldsmith, best known for his gently satirical novel *The Vicar of Wakefield* (1766), latterly wrote an immensely popular text on natural history entitled *A History of the Earth, and Animated Nature* (1774) which did much to awaken interest in African travel. Early travellers and explorers to the continent captured their impressions in Picturesque and Romantic terms with descriptions of Lake Victoria being indistinguishable from Romantic evaluations of Lake Windermere. David Livingstone's missionary travels and research in South Africa (1857) brought him celebrity status and many others published accounts of their adventures. Smith (2013) suggests that 'to the audience first encountering Africa through these texts, it could be argued that the sheer "otherworldliness" of the content must have read much like fiction' (p. 147). The landscapes, the climate, the wildlife, and the people were all incorporated into these tales which evoked danger, adventure and were consumed as a form of African Gothic which then set the tone for tourism on the continent for years to come.

Box 2.1 Case study: Historical development of literary tourism in China

China was the fourth most popular destination for international visitors in 2019, attracting 65.7 million tourists (United Nations World Tourism Organisation, UNWTO, 2021). This performance is even more impressive when one considers that China was closed to all international visitors during the People's Republic (1949–1974) with tourism development only being introduced in the late 1970s to promote economic growth.

China's literary culture dates back several thousand years but the development of Chinese literary tourism has a much more recent history. Chinese literary tourism resources now include writers' homes, monuments, constructed literary locations, and literary theme parks.

Wang and Zhang (2017) present a timeline for the establishment of writers' house museums in China, starting with the first, Lu Xun's home in Shanghai which first opened in 1951, with further homes of the author being added in 1953 and 59 (see Fig. 4.1). Lu Xun (1881–1936) was an influential author and his homes in Shanghai, Shaoxing, and Guangzhou still attract many visitors. The next stage of development was in the 1980s with nine writers' houses being added to the list, then three in the 1990s. The majority of these literary houses (18 sites) were founded in the 2000s, no doubt as a response to the 1999 Golden Week holiday policy introduced to stimulate tourism growth.

Other literary tourism sites were developed as follows:

Grand View Gardens in Beijing were created in 1984–1989. The large site of over 32 acres included gardens and full-sized permanent reconstructions of palaces and pavilions represented in the famous eighteenth century Chinese novel *The Dream of the Red Mansion* by Cao Xueqin. These buildings form a film set for the television adaptation of the novel and the whole site is a popular literary-themed visitor attraction (Ryan *et al*., 2009)

Yueyang Tower in the city of Yueyang, constructed during the Tang Dynasty (716 AD), became popular as a scenic tower and was visited by officials and poets who left many poems which are now on display on tablets around the site. The tower became widely known due to the writer Fan Zhongyan's *Remarks of Yueyang Tower* (1046), which provide a vivid description of the structure and surroundings on the shores of Dongting Lake. Yueyang Tower's fame has remained, and the building and its surroundings became a National Scenic Area in 1988 (Yu and Xu, 2018).

San Mao Teahouse in Zhouzhang opened in the early 1990s and is a site dedicated to the memory of celebrated Chinese author, San Mao (1943–1991). San Mao, a life-long traveller, had decided to settle in the historic town of Zhouzhang but died before realising her ambition. A local writer opened the teahouse in her memory and literary fans visit and leave their own writings behind. Themed walking routes have also been developed in the area exploring places significant to San Mao (Jiang and Xu 2017).

Literary tourism sites continue to be developed, one of the most significant being a planned recreation of Shakespeare's birthplace, authenticated by the Shakespeare Birthplace Trust in the UK. This faithful copy of the original building will be built in the town of Sanweng near Fuzhou. Other writers celebrated in Sanweng include Cervantes and Tang Xianzu (Joubin, 2019; Shakespeare Birthplace Trust, 2023).

Nineteenth century travel and the practice of reading

In order to understand the continuing development of literary tourism into the nineteenth century, it is important to acknowledge the enabling factors that made such travel possible – namely transport, developments in publishing, and the evolution of a literate travelling public. Travel up until the mid-nineteenth century was for the elite, who had their own equipage, or the less wealthy who could afford to book a seat on a stagecoach. These journeys would have been undertaken with some trepidation, mindful of highwaymen and cut-purses who infested the unpoliced routes. Longer journeys necessitated overnight stays at coaching inns to refresh or change one's horses or, for the better-connected traveller, stopping off at the country residences of members of their wider social circle. Steamships allowed wealthier travellers to go further afield although this would not have been considered a comfortable mode of transport as many accounts show. Travel for the British masses was not possible until the coming of passenger railway services in the 1830s which offered inexpensive connections to seaside resorts and, for the middle class would-be Grand Tourist, easy access to the coast for European travel or the pleasures of the Lake District. For the more adventurous, the railways also opened up access to the Highlands of Scotland and to Cornwall, the last English county to be connected to the railway network in 1859, before which it was believed that more English visitors had been to Paris than to Truro (Simmons, 1984, p. 213). Comfortable railway termini hotels owned or leased by the railway companies sprang up in cities such as London, York, and Edinburgh and at the ports of Folkstone, Harwich, and Holyhead. These were considered to be eminently respectable establishments because of the connection with the railway companies and were the first hotels to offer public rooms for ladies' use, thus opening up safe spaces for female travellers.

In tandem with the greater ease and accessibility to travel came increasing levels of literacy and availability of printed matter. Global literacy doubled in the nineteenth century and, across Europe, the Enlightenment had left a legacy of increased reading ability, particularly in the Northwest. In England, two-thirds of men and about half of women were literate in 1840, rising and equalising, thanks to educational reform, to 97% of all adults by the end of the nineteenth century (Lloyd, 2007). From the seventeenth to nineteenth centuries, the working classes would have read broadsides which were single sheets of printed paper sold on the streets for a half penny and conveying official announcements, news, songs, and ballads. However, this reading matter was soon to be extended as improvements in access to education and the steep rises in literacy levels described above meant that reading became a popular pastime across all social classes and reading matter was priced to be affordable for most would-be readers. Serialisation was a popular and affordable publishing method and some of Charles Dickens' and George Eliot's work was first produced as monthly magazine serials, with exciting cliff-hanger endings being used to entice the reader back for the next instalment. Novels were also published in a three-volume format but as these were extremely expensive (they would routinely cost the same as an average weekly industrial wage) they were usually borrowed from libraries by all strata of society (Flint, 2014). These commercial circulating libraries would deliver the latest publications to their subscribers and delivered boxes of books across the country to their borrowers. Mudie's was the most famous of these libraries and was set up in 1842, offering a 'select' collection of three-volume ('triple-decker') novels for an annual subscription rate of one guinea. Later editions were then much more cheaply produced as single volumes or 'railway editions,' so called because they were sold at stations and read by the travelling public. Lending libraries were also to be found at railway stations, underpinning the close connection between the development of a reading public and the railway network. Books were delivered by rail, sold and lent at stations and, importantly, passengers used their travelling time to read (Stephens, 1990). In England, the Public Libraries Act of 1850 gave local authorities permission to set up public libraries although in fact few were built until Queen Victoria's Golden Jubilee in 1887. Prior to this, private philanthropists such as the Scottish-American steel magnate, Andrew Carnegie, provided small and medium sized libraries across the country and, by 1914, 62% of England's population could access a local free library (Historic England, 2016).

Although sustained works of storytelling have existed throughout history in the form of sagas and romances, the novel, which we now consider to be the mainstay of literary fiction output and consumption, is often said to date back to the eighteenth century, with Daniel Defoe's *Robinson Crusoe* (1719) and *Moll Flanders* (1722) and Samuel Richardson's *Pamela* (1740) being early examples of the genre of book-length works of fiction (Moore, 2013). These novels featured strong central characters battling with moral questions and reading these contemporary works was often considered to be inferior to the study of classical literature, history, and poetry. The nineteenth century was, however, the great era of the novel, and these proliferated during the Victorian period, first influenced by Romanticism which produced idealised characters and situations and then by a very Victorian zeal for the improvement of society, the education of readers, and the raising of social issues. This resulted in realistic novels with characters who were neither wholly good nor evil and one of the most successful proponents of this style of writing was Charles Dickens, who managed to combine all these worthy goals with hugely entertaining and memorable characters and storylines. Similarly, writers such as Elizabeth Gaskell and George Eliot were praised for their accurate, politically- and psychologically-informed works depicting urban and rural life, respectively. These novels were read aloud in family parlours, or alone in railway carriages, discussed at tea and in the burgeoning book groups of the time. Reading for educational purposes was much approved of within Victorian society and the move towards purposeful 'rational recreation' meant that the passivity and idleness that reading may inculcate must be addressed by the rigorous practice of note-taking and the copying of quotations, all of which was a common way of reading, sharing, and taking ownership of a work of fiction which would more often than not have to be returned to the circulating library (Price, 2012).

A further development that facilitated the spread of reading, and hence literary tourism, was the arrival of gas and electric lighting in homes allowing readers to continue their perusals after dark (Flint, 2014). For centuries, poor people would construct rushlights by coating dried rushes with layers of fat before burning. Tallow candles lasted longer but smoked and smelled strongly of animal fat. The wealthy could afford the much more fragrant beeswax and heightened the effect with mirrored walls which amplified some of the light but reading by candlelight must have taken its toll on eyesight. Oil lamps began to replace candles in the later eighteenth century and, by the 1840s, town dwellers were gradually able to have gas lighting. The very wealthy who could afford electric light bulbs and their own generators could also enter the exciting world of electrified lighting in the 1880s, but it wasn't until the National Grid was created in the UK in the 1930s that electricity found its way into the homes of the masses (Dillon, 2002), thus facilitating many more hours of reading and other home-based leisure pursuits which were both in direct competition with, and an accompaniment to, reading for pleasure as we will see later.

Literature and travel in the twentieth century

Many of the visiting trends developed in the Victorian era continued into the new century with the homes of authors continuing to attract visitors in increasing numbers. Butler describes this continuity and 'the links between literature and the establishment of nineteenth and early twentieth century tourist destinations in determining a pattern still prevalent today' (Butler, 1990, p. 46). However, as Robinson and Andersen note, visiting literary sites as a touristic activity was still on a very small scale at this point and limited mainly to 'middle-class enthusiasts and emergent literary societies' (2002b, p. 13). At the start of the twentieth century, visiting the homes of authors was probably the most significant form of organised literary tourism and there was already a number of these sites available to visitors with houses associated with Walter Scott, William Shakespeare, Robert Burns, Thomas Carlyle, John Milton, and William Wordsworth all having opened during the previous century. Nonetheless, according to Young (2015), within the UK and further afield, the majority of the authors' homes currently operating as visitor attractions were acquired and opened up to the public in the last century (see Table 2.1) usually through the efforts of

Table 2.1 Writers' homes and the year of their public opening

Writer's home	*Year of public opening*
Robert Burns birthplace, Ayrshire, Scotland	1881
Thomas Carlyle's house, London, England	1895
Keats-Shelley House, Rome, Italy (See Fig. 2.1)	1909
L.M. Alcott's Orchard House, Concorde, Massachusetts, United States of America (USA)	1912
Sándor Petőfi House, Kiskőrös, Hungary	1951
Miguel de Cervantes Birthplace Museum, Alcalá de Henares, Spain	1956
Ernest Hemingway's house, Key West, Florida, USA	1964
Karen Blixen Museum, Nairobi, Kenya	1986
Colleen McCullough's house, Norfolk Island, Australia	2015

voluntary committees and, latterly, local government. The power and significance of the authorial home will be analysed in detail in Chapter Four but its role in the historic evolution of fiction-based tourism is useful to consider here.

Writing in an article entitled 'Great Men's Houses,' published in *Good Housekeeping* in March 1932, Virginia Woolf comments on the ability of the author's home to conjure up a wealth of both overt and implied information for the visitor. In her article, Woolf describes her visit to the relatively modest home of Thomas and Jane Carlyle in Cheyne Row in Chelsea, London which first opened its door to the public in the 1890s, stating that the house and its contents – and particularly its spartan plumbing arrangements – tells one more about the couple's everyday life and the context for their work than a string of biographies. The collection of nibs, pen holders, and items of clothing on display added to the domestic scene and Woolf concluded that life in a more well-appointed house would have been much easier for the Carlyles but perhaps it was the struggle that made them who they were. Such are the reflections that authors' homes inspire in their visitors (Woolf, 1980).

The shaping of twentieth century European tourism

Literary fiction was also playing a more generalised role in popularising specific tourist destinations and influencing holiday-making practices. This was particularly true in the shaping of the literary landscape of the Mediterranean

Figure 2.1 Keats-Shelley House in Rome which opened to the public in 1909. Photograph by J. Kennell.

for American and British visitors. American author F. Scott Fitzgerald's own lifestyle and that portrayed in his novel *Tender is the Night* (1934) did much to consolidate in readers' minds the French Riviera as a glamorous holiday destination. The Riviera had been considered a healthy winter resort for the upper classes for some time as the mild climate was thought beneficial for curing tuberculosis (Keller, 2010). The lack of air conditioning and refrigeration meant that were no international visitors in the warmer months at the start of the twentieth century, but Fitzgerald was influential in shifting visiting patterns to the summer and moving holiday activities to the beach. In particular, he popularised and made socially-acceptable the business of acquiring a suntan. His 'representations of high-class cosmopolitan society … helped transform beaches and suntanning from a rare and shocking symbol of the lower classes to an eminently fashionable trend' (Keller, 2010, p. 131) which continued well into the late twentieth century before concerns about the link between sunburn and skin cancer significantly lessened its hold. However, the glamour of the beaches of the South of France still holds sway and St Tropez self-tanning products, launched in 1993, continue to allow users to safely sport the sun-kissed Riviera look (Chesters, 2012).

Ernest Hemingway's travels in Spain also popularised the culture of this country for a wide readership, in particular his meticulous and accurate detailing of the Running of the Bulls event at the San Fermin festival in Pamplona which he attended over several years and documented in his novel *The Sun Also Rises/Fiesta*, published in 1926 (Krauth, 2019). Hemingway's close association with Spain was further cemented by the publication of *For Whom the Bell Tolls* (1940) and it is now possible to 'holiday in and explore … Hemingway's Spain' (Robinson and Anderson 2002b, p. 21). These American writers wielded much influence across Europe but similarly the 'British literary diaspora in the Mediterranean' (Moyà Antón, 2011) was helping to construct an image of this area which had previously been shaped by Hellenic and classical discourse but was, in the early decades of the twentieth century, being associated with exoticism and freedom, encapsulated in olive groves, citrus trees, inexpensive wine, and a slower pace of life which was in heady contrast to the perceived constraints of the British way of life. The Bohemian lifestyle of writer Robert Graves, who spent most of his adult life living and working on the Balearic Island of Mallorca between 1929 and 1985, brought a similarly hedonistic type of visitor to the small coastal village of Deia in the north of the island (Barke, 2002) and his former home still features in coach tours of the area. Other members of this diaspora also popularised regions of the Mediterranean with both their work and their adopted lifestyles. Norman Douglas settled in Capri and encapsulated it in *South Wind* (1917), a fictionalised account of island life that was irresistible to would-be visitors. In the Ionian region, Lawrence Durrell did the same for Corfu, along with his younger brother Gerald's autobiographical tales of his family and island wildlife (see Box 9.1). Remnants of these writers' lives can still be found across the Mediterranean and their influence lingers despite the widespread development of mass tourism in the region.

Mapping literary tourism

The close link between fiction and place was furthered in the twentieth century by the continued production of dedicated literary atlases and gazettes which were, as Watson (2009) suggests, a culmination of all the travelogues, diaries, literary itineraries, and homes and haunts guide books which had been produced by and for travellers from the early days of the Grand Tour onwards. These literary guidebooks help to create tourist destinations from sites of literary interest in that they 'classify the authors and places deemed worthy of the "tourist gaze" and point the literary tourist towards them' (Philips, 2011, p. 22). These literary guides, such as Brandt's *Literary Landscapes* published in 1951, *The Oxford Literary Guide to the British Isles* (1977 and now in its third edition), and Daiches and Flower's *Literary landscapes of the British Isles: A Narrative Atlas* (1979), are complete with photographs, illustrations, and maps to guide the would-be literary tourist to the places which will most promote a greater appreciation of the literary work. Daiches and Flower (1979) are keen to point out in their introduction that their Atlas, with its historical maps and black and white illustrations, is 'concerned with new perceptions into the life of the literary work rather than with what might be called the tourist's view of literary shrines' (p. 7). However much these rather scholarly guides may wish to distance themselves from any whiff of authorial idolatry, they were highly influential in dictating the canon of literary sites deemed worthy of visitation. In her analysis of these guides Philips (2011) notes that they are highly selective (they all include Dickens, Walter Scott, Shakespeare, the Brontës) and nostalgic for the rural past. This becomes clear when even very urban-based authors such as Dickens are associated with rural landscapes, in his case the county of Kent rather than the streets of London.

Broadcast media and literary themes

Guides such as these appealed to the scholarly visitor and the armchair traveller but the still relatively small numbers of literary visitors in the first half of the twentieth century were to be augmented by the introduction of mass media which would open up the markets for literary output and increase the interest in literary-inspired travel. The radio was first patented in 1900 and brought news, music, and literary output into homes. Scenes from Jane Austen's *Pride and Prejudice* were broadcast in 1924 on '2L0,' the British Broadcasting Corporation's (BBC) first radio station, heard only in the London area, whilst the works of Dickens, Shakespeare, and Conan Doyle could be heard on early American broadcast networks (BBC, 2023). In the UK, the BBC started television broadcasts in 1932 and some early literary programming included the first televised Shakespeare play, *Twelfth Night*, broadcast in 1937, with an adaptation of Dickens' *A Christmas Carol* following in 1943 (BBC Archive, 2023). By 1955, television had made its way into the living rooms of half of the American population. In the same year, transmitter coverage meant that 95%

of the UK had BBC television reception (National Science and Media Museum, 2011). High quality televised literary adaptations have continued to be a mainstay of radio and television output, reaching large audiences, and providing a significant impetus for literary tourism as we will see in Chapter Nine. Many literary sites embrace these developments by incorporating sets, costumes, and props from recent film or television adaptations into their exhibitions and selling DVDs in their gift shops to extend the appeal and augment the imaginative power of their visitors.

Destination marketing – selling the literary tourism country

As interest in places with literary associations increased in the twentieth century, in tandem with the development of the wider cultural and heritage tourism market (see Chapter Three), the accompanying commercialisation of literary tourism became evident. As Hoppen, Brown and Fyall (2014, p. 43) state in their discussion of the uses of literary associations in the branding of destinations, 'an increasing number of private and public sector agencies in the UK have realized the potential of literary tourism and are promoting specific images of those places.' Thus, agencies such as destination management organisations and national tourist boards began to see the value of promoting the cultural and literary aspects of their regions as a way of differentiating and providing a brand identity to lure both domestic and international visitors. In the early 1980s, place marketing became a strategic development policy across Europe and North America in areas of industrial decline and specific cultural features such as links to authors of popular literary works were exploited. The creation of literary 'countries' was already well-established in the UK with Shakespeare, Scott, Burns, and Hardy countries all mapped out in book format in the early guides discussed above, and Wordsworth country in the Lake District being described as paradigmatic of this way of conceptualising place for the well-read visitor (Hess, 2016). However, the deliberate creation of literary-themed 'countries' for the purposes of tourism promotion was a newer approach within wider destination marketing, for example the creation of Catherine Cookson Country which was launched in 1985 in South Tyneside in the Northeast of England (Pocock, 1992). This literary country was produced around the life and works of the prolific and very popular twentieth century author who set her social history novels in the region in the period between 1850s and 1930s. The deliberate designation of this country was an attempt to replace industrial decline with a tourism-oriented product created around the heritage and traditions of the region.

Similar 'countries' can be found around the UK and Ireland such as Yeats Country in County Sligo, Ireland, and Dickens Country in Kent, England, easily identifiable by their brown signs which indicate places of touristic interest. In the United States, more recent literary country development has been introduced to address the aftermath of the economic recession. Steinbeck Country, exploring the life and work of iconic American author and

Nobel Laureate, John Steinbeck, in Salinas, California grew out of the creation of the Salinas Valley Tourism and Visitors' Bureau in 2008 (SVTVB, 2021). Steinbeck fans can follow one of four Steinbeck trails, visit the National Steinbeck Centre, appraise literary locations, and even take lunch in his boyhood home which now operates a restaurant specialising in locally sourced produce (see Fig. 1.1). However, not all literary themes are grasped as enthusiastically by tourism planners, as Light (2008, 2017) explains in the case of Romania. The literary place myth that has been created for Transylvania is not one that is recognised by the local population and consequently, its Dracula connections are not promoted, despite the high levels of interest from American and British visitors who still arrive either independently or using the services of overseas tour operators.

In addition to the development of literary countries, destination management organisations also began to emulate the literary maps published in guides such as those discussed above. Rather than being adjuncts to the experience of reading, these maps were resolutely intended as promotional tools to encourage visitation to destinations with literary associations. This would both benefit the individual locations who might receive additional footfall from the literary-inspired tourist and create a favourable impression of the region as one with a strong literary heritage. The British Tourist Authority (BTA) published a map of *Literary Britain* in 1998 which featured 82 authors and their landscapes including the usual canonical Shakespeare, Dickens, Austen, and Wordsworth but also featuring James Herriot's Yorkshire and Daphne du Maurier's Cornwall. The inclusion of these more populist authors was undoubtedly due to their links with very well-received television adaptations of their works and the fact that they were already associated with the traditional holiday regions of the Yorkshire Dales and Cornwall's coast and moors. The BTA launched a newer version of the map in 2011 – this also had 82 authors, but this time included three writers who were being utilised to boost tourism to the rather under-visited West Midlands, namely Jerome K. Jerome, J.R.R. Tolkien and George Eliot (Calder, 2011). The newest incarnation of the map was launched in 2017 to celebrate Visit England's Year of Literary Heroes – an event underpinned by the 200-year anniversary of Jane Austen's death and the twentieth anniversary of the first Harry Potter novel.

Literary tourism continues to develop in terms of visitor interest and in the range of products and experiences available. As we shall see in subsequent chapters, contemporary literary tourism now makes full use of digital media to explore time-honoured themes, with mobile applications providing audio and visual tours and interactive maps, Quick-Response (QR) readers offering *in-situ* information in literary destinations, and agencies such as the National Trust in the UK producing podcasts exploring the literary themes of properties in their care. The medium may change but the message remains the same – there is an abiding interest in the writers of creative works, their fictional output, and the ability to locate these within spaces which can be toured.

Conclusion

In this chapter we have traced the historic evolution of literary tourism from its earliest Classical roots to its contemporary landscapes, noting the various influences underpinning its development. In particular we have seen that travel associated with literary themes has mirrored tourism in general in its changing focus from the individual pursuit of author-oriented knowledge to a much more widely consumed and diffuse experience of themes and locations. The emergence of an increasingly literate travelling public has also given a major impetus to the evolution of literary fiction-inspired tourism, a momentum that has only increased with the extension of print into broadcast media. In the following chapter we will situate the study of fiction-inspired travel within the wider context of cultural and heritage tourism and explore a range of theoretical perspectives through which this activity can be studied.

References

Abbas, E. (2013) 'The tale of the shipwrecked sailor and the rite of passage', *Journal of the American Research Center in Egypt*, pp. 9–18. Available at: https://www.jstor.org/stable/24555414 (Accessed: 29 June 2022).

Adler, J. (1989) 'Origins of sightseeing', *Annals of Tourism Research*, 16 (1), pp. 7–29. Available at: https://doi.org/10.1016/0160-7383(89)90028–90025 (Accessed: 29 June 2022).

Aitchison, C., MacLeod, N.E. and Shaw, S.J. (2002) *Leisure and tourism landscapes: Social and cultural geographies.* Abingdon: Routledge.

Barke, M. (2002) '"Inside" and "outside" writings on Spain: Their relationship to Spanish tourism', in Robinson, M. and Andersen, H.C. (eds.) *Literature and tourism: Reading and writing tourism texts.* London: Continuum, pp. 80–104.

Barnard, R. (2002) 'Tourism comes to Haworth', in Robinson, M. and Andersen, H.C. (eds.) *Literature and tourism: Reading and writing tourism texts.* London: Continuum, pp. 143–154.

BBC (2023) History of the BBC: Jane Austen at the BBC. Available at: https://www.bbc.com/historyofthebbc/research/jane-austen-at-the-bbc (Accessed: 12 June 2023).

BBC Archive (2023) Programme index. Available at: https://genome.ch.bbc.co.uk/ (Accessed: 12 June 2023).

Bhandari, K. (2008) 'Touristification of cultural resources: A case study of Robert Burns', *Tourism: An International Interdisciplinary Journal*, 56 (3), pp. 283–293. Available at: https://hrcak.srce.hr/36621 (Accessed: 13 February 2022).

Black, J. (2010) *The British and the Grand Tour* (Routledge Revivals). First published 1985. Abingdon: Routledge.

Booth, A. (2016) *Homes and haunts: Touring writers' shrines and countries.* Oxford: Oxford University Press.

Buchmann, A. (2006) 'From Erewhon to Edoras: Tourism and myths in New Zealand', *Tourism Culture & Communication*, 6 (3), pp. 181–189. Available at: https://doi.org/10.3727/109830406778134090 (Accessed: 15 February 2022).

Buell, L. (1989) 'The Thoreauvian pilgrimage: The structure of an American cult', *American Literature*, 61 (2), pp. 175–199. Available at: https://doi.org/10.2307/2926692 (Accessed: 5 January 2022).

Butler, R.W. (1990) 'The influence of the media in shaping international tourist patterns', *Tourism Recreation Research*, 15 (2), pp. 46–53. Available at: https://doi.org/10.1080/02508281.1990.11014584 (Accessed: 27 June 2023).

Buzard, J. (1992) *The beaten track: European tourism, literature, and the ways to culture, 1800–1918*. Oxford: Oxford University Press.

Calder, S. (2011) 'From Cornwall to Braemar, Du Maurier to Wesley, literary Britain is on the map', *The Independent*, 22 September. Available at: https://www.independent.co.uk/news/uk/this-britain/from-cornwall-to-braemar-du-maurier-to-wesley-literary-britain-is-on-the-map-698783.html (Accessed: 22 March 2021).

Chesters, A. (2012) 'A brief history of St Tropez' *The Guardian*, 10 April. Available at: https://www.theguardian.com/fashion/fashion-blog/2012/apr/10/brief-history-of-st-tropez (Accessed: 13 December 2021).

Daiches, D. and Flower, J. (1979) *Literary landscapes of the British Isles: A narrative atlas*. London: Paddington Press.

Dillon, M. (2002) *Artificial sunshine: A social history of domestic lighting*. Swindon: National Trust.

Donaldson, C., Gregory, I.N. and Murrieta-Flores, P. (2015) 'Mapping "Wordsworthshire": A GIS study of literary tourism in Victorian Lakeland', *Journal of Victorian Culture*, 20 (3), pp. 287–307. Available at: https://doi.org/10.1080/13555502.2015.1058089 (Accessed: 12 June 2023).

Feifer, M. (1985) *Going places: The ways of the tourist from Imperial Rome to the present day*. London: MacMillan.

Flint, K. (2014) 'Victorian readers', London: British Library. Available at: https://www.bl.uk/romantics-and-victorians/articles/victorian-readers (Accessed: 16 March 2021).

Foster, S. (2009) 'Americans and anti-tourism', in Watson, N.J. (ed.) *Literary tourism and nineteenth-century culture*. Basingstoke: Palgrave MacMillan, pp. 175–183.

Gold, J.R. and Gold, M.M. (1995) *Imagining Scotland: Tradition, representation, and promotion in Scottish tourism since 1750*. Aldershot: Scolar Press.

Goodman, T.M. (2019) '*"Prized by the tourist": Souvenir books in Victorian Scotland*', PhD thesis, University of Edinburgh, Edinburgh. Available at: https://era.ed.ac.uk/handle/1842/36884 (Accessed: 30 June 2023).

Hazard, E. (2016) 'The attraction of the American author's house: Architecture, guidebooks, and display, 1853–1904', *Art History: Journal of the Association of Art Historians*, 39 (5), pp. 896–925.

Hendrix, H. (2009) 'From early modern to Romantic literary tourism: A diachronical perspective', in Watson, N.J. (ed.) *Literary tourism and nineteenth-century culture*. London: Palgrave Macmillan, pp. 13–24.

Herbert, D. (2001) 'Literary places, tourism and the heritage experience', *Annals of Tourism Research*, 28 (2), pp. 312–333. Available at: https://doi.org/10.1016/S0160-7383(00)00048–00047 (Accessed: 3 November 2021).

Hess, S. (2016) 'Wordsworthshire and Thoreau country: Transatlantic landscapes of genius', in Westover, P. and Rowland, A.W. (eds.) *Transatlantic literature and author love in the nineteenth century*. London: Palgrave MacMillan, pp. 175–201.

Historic England (2016) *The English public library 1850–1939*. London: Historic England.

Hoppen, A., Brown, L. and Fyall, A. (2014) 'Literary tourism: Opportunities and challenges for the marketing and branding of destinations?', *Journal of Destination Marketing & Management*, 3 (1), pp. 37–47. Available at: https://doi.org/10.1016/j.jdmm.2013.12.009 (Accessed: 2 February 2021).

Howitt, W. (1875) *Homes and haunts of the most eminent British poets.* 3rd edn. London: Routledge.

Hulme, P.D. and Youngs, T. (2002) 'Introduction' in Hulme, P.D. and Youngs, T. (eds.) *The Cambridge companion to travel writing*. Cambridge: Cambridge University Press, pp. 1–14.

Jiang, L. and Xu, H. (2017) 'The growth of literary places in ancient town tourism destinations: Based on the theories of Bourdieu, Danto and Dickie', *Journal of Tourism and Cultural Change*, 15 (3), pp. 213–228. Available at: https://doi.org/10.1080/14766825.2016.1148153 (Accessed: 10 February 2021).

Joubin, A.A. (2019) 'Performing commemoration: The cultural politics of locating Tang Xianzu and Shakespeare', *Asian Theatre Journal*, 36 (2), pp. 275–280. Available at: https://www.jstor.org/stable/26784363 (Accessed: 15 March 2021).

Keller, S.L. (2010) 'The Riviera's golden boy: Fitzgerald, cosmopolitan tanning, and racial commodities in "Tender Is the Night"', *The F. Scott Fitzgerald Review*, 8, pp. 130–159. Available at: https://www.jstor.org/stable/41583159 (Accessed: 20 March 2021).

Krauth, N. (2019) 'Literary tourism: Readers, writers and being there', *TEXT*, 23 (56), pp. 1–15.

Light, D. (2008) 'Imaginative geographies, Dracula and the Transylvania place myth', *Human Geographies*, 2 (2), pp. 6–17. Available at: http://www.gabrielsimion.net/humangeographies.org.ro/articles/22/Light.pdf (Accessed: 12 July 2023).

Light, D. (2017) 'The undead and dark tourism: Dracula tourism in Romania', in Hooper, G. and Lennon, J.J. (eds.) *Dark tourism: Practice and interpretation*. Abingdon: Routledge, pp. 121–133.

Lloyd, A.J. (2007) 'Education, literacy and the reading public', *British Library Newspapers*. Available at: https://www.gale.com/intl/essays/amy-j-lloyd-education-literacy-reading-public (Accessed: 30 June 2023).

Mansfield, C. (2015) *Researching literary tourism*. Bideford: Shadows Books.

Moore, S. (2013) *The novel: An alternative history, 1600–1800*. London: Bloomsbury Publishing.

Moyà Antón, E. (2011) 'British literary diaspora in the Mediterranean: The (re) creation of the "Sunny South"', *Kaleidoscope*, 5 (1), pp. 33–44.

National Science and Media Museum (2011) A short history of British television. Available at: https://blog.scienceandmediamuseum.org.uk/history-of-british-television-timeline (Accessed: 2 June 2023).

Nestor, M.C. (2020) 'Revisiting the Lady of the Lake: Walter Scott and the representation of Scotland', *Scottish Literary Review*, 12 (1), pp. 61–80. Available at: https://muse.jhu.edu/article/757382 (Accessed: 25 March 2021).

Philips, D. (2011) 'Mapping literary Britain: Tourist guides to literary landscapes 1951–2007', *Tourist Studies*, 11 (1), pp. 21–35. Available at: doi:10.1177/1468797611412014.

Pocock, D. (1992) 'Catherine Cookson country: Tourist expectation and experience', *Geography*, 77 (3), pp. 236–243. Available at: https://www.jstor.org/stable/40572194 (Accessed: 3 April 2021).

Price, L. (2012) 'Victorian reading', in Flint, K. (ed.) *The Cambridge history of Victorian literature*. Cambridge: Cambridge University Press, pp. 34–55.

Punter, D. (1994) 'The picturesque and the sublime: Two worldscapes', in Copely, S. and Garside, P. (eds.) *The politics of the picturesque: Literature, landscape and aesthetics since 1770*. Cambridge: Cambridge University Press, pp. 220–239.

Quinteiro, S. and Baleiro, R. (2018) *Key concepts in literature and tourism studies.* Lisbon: University of Lisbon.

Roberts, D. (2009) 'Harriet Beecher Stowe and Florida tourism', in Watson, N.J. (ed.) *Literary tourism and nineteenth century culture.* Basingstoke: Palgrave MacMillan, pp. 196–209.

Robinson, M. and Andersen, H.C. (2002a) 'Introduction', in Robinson, M. and Andersen, H.C. (eds.) *Literature and tourism: Reading and writing tourism texts.* London: Continuum, pp. xiii–xix.

Robinson, M. and Andersen, H.C. (2002b) 'Reading between the lines: Literature and the creation of touristic spaces' in Robinson, M. and Andersen, H.C. (eds.) *Literature and tourism: Reading and writing tourism texts.* London: Continuum, pp. 1–38.

Ryan, C., Yanning, Z., Huimin, G. and Song, L. (2009) 'Tourism, a classic novel, and television: The case of Cáo Xuěqin's Dream of the Red Mansions and Grand View Gardens, Beijing', *Journal of Travel Research*, 48 (1), pp. 14–28. Available at: doi:10.1177/0047287508328796.

Santesso, A. (2004) 'The birth of the birthplace: Bread Street and literary tourism before Stratford', *ELH*, 71 (2), pp. 377–403. Available at: https://www.jstor.org/stable/30030055 (Accessed: 20 March 2021).

Schaff, B. (2009) 'John Murray's handbooks to Italy: Making tourism literary', in Watson, N.J. (ed.) *Literary tourism and nineteenth century culture.* Basingstoke: Palgrave MacMillan, pp. 106–118.

Shakespeare Birthplace Trust (2023) The Birthplace now. Available at: https://www.shakespeare.org.uk/explore-shakespeare/shakespedia/shakespeares-birthplace/birthplace-now/ (Accessed: 30 June 2023).

Simmons, J. (1984) 'Railways, hotels, and tourism in Great Britain 1839–1914', *Journal of Contemporary History*, 19 (2), pp. 201–222. Available at: https://doi.org/10.1177/002200948401900203 (Accessed: 22 March 2021).

Smith, Y. (2013) 'From the Grand Tour to African adventure: Haggard-inspired literary tourism', *South African Journal of Cultural History*, 27 (2), pp. 132–156. Available at: https://hdl.handle.net/10520/EJC148089 (Accessed: 30 March 2021).

Stephens, W.B. (1990) 'Literacy in England, Scotland, and Wales, 1500–1900', *History of Education Quarterly*, 30 (4) pp. 545–571 Available at: https://doi.org/10.2307/368946 (Accessed: 30 March 2021).

SVTVB (2021) Salinas Valley Tourism and Visitors Bureau – Central Coast Soul of California. Available at: https://soulofca.org/about/contact/ (Accessed: 18 March 2021).

Towner, J. (1985) 'The Grand Tour: A key phase in the history of tourism', *Annals of Tourism Research*, 12 (3), pp. 297–333. Available at: https://doi.org/10.1016/0160-7383(85)90002–90007 (Accessed: 20 June 2023).

Towner, J. (2002) 'Literature, tourism, and the Grand Tour', in Robinson, M. and Andersen, H.C. (eds.) *Literature and tourism: Reading and writing tourism texts.* London: Continuum, pp. 226–238.

UNWTO (2021) *International tourism highlights, 2020 edition.* Madrid: UNWTO. Available at: https://doi.org/10.18111/9789284422456 (Accessed: 20 March 2022).

Vogel, A. (2002) 'African sublime: The Dark Continent joins the Grand Tour', in Robinson, M. and Andersen, H.C. (eds.) *Literature and tourism: Reading and writing tourism texts.* London: Continuum, pp. 279–295.

Wang, H.J. and Zhang, D. (2017) 'Comparing literary tourism in mainland China and Taiwan: The Lu Xun native place and the Lin Yutang house', *Tourism*

Management, 59, pp. 234–253 Available at: https://doi.org/10.1016/j.tourman.2016.08.008 (Accessed: 30 March 2021).

Watson, N.J. (2006) *The literary tourist: Readers and places in Romantic and Victorian Britain*. Basingstoke: Palgrave MacMillan.

Watson, N.J. (2009) 'Introduction', in Watson, N.J. (ed.) *Literary tourism and nineteenth-century culture*. Basingstoke: Palgrave MacMillan, pp. 1–12.

Westaway, J. (2016) 'The origins and development of mountaineering and rock climbing tourism in the Lake District, c. 1800–1914', in Walton, J.K. and Wood, J. (eds.) *The making of a cultural landscape: The English Lake District as tourist destination 1750–2010*. London: Routledge, pp. 155–180.

Westover, P. (2009) 'How America "inherited" literary tourism', in Watson, N.J. (ed.) *Literary tourism and nineteenth-century culture*. Basingstoke: Palgrave MacMillan, pp. 184–195.

Woolf, V. (1980) *The London scene: The essays*. First published in 1932. London: Hogarth Press.

York Festival Trust (2023) York Festival Trust – Guardians of the Mystery Plays. Available at: https://www.yorkmysteryplays.co.uk/york-festival-trust/ (Accessed: 12 August 2021).

Young, L. (2015) 'Literature, museums, and national identity; or, why are there so many writers' house museums in Britain?', *Museum History Journal*, 8 (2), pp. 229–246. Available at: https://doi.org/10.1179/1936981615Z.00000000052 (Accessed: 13 April 2023).

Young, L. (2019) 'Writers' house museums: English literature in the heart and on the ground', in McCracken Fletcher, L. (ed.), *Literary tourism and the British Isles: History, imagination and the politics of the past*. Lanham MD: Lexington Books, pp. 165–190.

Yu, X. and Xu, H. (2018). 'Moral gaze at literary places: Experiencing "being the first to worry and the last to enjoy" at Yueyang Tower in China', *Tourism Management*, 65, pp. 292–302. Available at: https://doi.org/10.1016/j.tourman.2017.10.017 (Accessed: 13 April 2023).

3 Literary fiction tourism in context

As we have seen in the preceding chapters, literature has very close connections to geographical place and so it is not surprising that it has inspired a wide range of different touristic practices which are themselves engagements with physical settings. Throughout these pages, we will explore the wide range of locations, buildings, monuments, institutions, landscapes, and townscapes which attract visitors because of their literary associations – from commemorative plaques to walking trails, from literary hotels to writers' graves. Literary tourists may wish to gaze on the desk upon which a favourite novel was written or look out at a view that would have been familiar to a treasured author. They might seek out landscapes described in detail in a much-loved text, armed with a dog-eared copy from which to find clues. They may be inspired to visit a museum dedicated to the life of an author and gain a greater understanding of the artist and their work through the amassing of biographical detail. They might feel moved to make a pilgrimage to the grave of an individual writer or to gain a sense of the literary reputation of a city through viewing statues or discovering local bookshops. In an attempt to revisit much-loved childhood stories, a visitor might explore a theme park or undertake a guided tour which takes them on a packaged journey through the life and work of a writer. Literary tourists may also enjoy a more interactive experience which could include attending literary events and re-enactments which might involve writing, dressing up, role-play, and other interactions with the written word and its producers. All these diverse locations and activities are inspired in some way or another by the lives and imaginative output of writers and are aspects of literary fiction tourism.

The activities and experiences outlined above utilise cultural resources, themes, and associations and the literary output of a region is part of its wider cultural tourism offering. Therefore, in this chapter, as an activity that draws on the life and creative works of artists, we situate the phenomena of literary-inspired tourism within the wider context of cultural tourism, exploring the development of culturally motivated travel and the changing nature of its production and consumption. We will then go on to explore a range of key theoretical perspectives which illuminate the study of both cultural tourism and literary fiction tourism. Each of these perspectives will be introduced briefly here and subsequently revisited and contextualised in the following chapters.

DOI: 10.4324/9781003041740-3

As we have seen in Chapter Two, the history of touring for pleasure has ancient roots but tourism as an activity for the masses is a relatively recent phenomenon, driven by increased leisure time, disposable income, and transport availability. Defined by the United Nations World Tourism Organisation (UNWTO, 2022) as a 'social, cultural, and economic phenomenon which entails the movement of people to countries or places outside their usual environment for personal or business/professional purposes,' tourism is now one of the largest economic sectors, incorporating more of the world's territories and populations every year, albeit with a hiatus in 2020–2022 due to the Covid-19 pandemic. Mass travel has long been associated with coastal and beach tourism – the traditional sun, sea, and sand holiday package – which encapsulates the hedonistic and recreational aspects of the tourist experience. Whilst these holidays remain very popular, with the Association of British Travel Agents stating that package holidays were the preferred choice of 51% of United Kingdom (UK) consumers in 2019 (ABTA, 2020, p. 4), there has been an increasing specialisation of the tourism product over the last few decades, with the creation of ever more individualised niches being part of the postmodern condition of the late twentieth century.

This shift away from the more traditional mass tourism products towards trips taken for more individualised leisure motivations has underpinned the important role of culture within tourism and emphasised its sociological nature. This has been clearly exemplified in the academic and managerial literature on tourism produced from the post-war era of the twentieth century onwards. Initially, these studies focussed on the logistics of getting visitors to destinations, taking care of them, and maximising the profit and efficiency of these processes. The experiences and motivations of the visitor themselves were assumed to be relaxation, escapism, and a chance to spend time with family and friends. Some of the earliest twentieth century studies of tourism took a rather dismissive view in contrasting the classical age of gentlemanly exploration with the mass tourism industry, for example in *The Lost Art of Travel* (1964), Daniel Boorstin characterised the mass tourist as one easily satisfied with pseudo-events and fake experiences. Similarly, Turner and Ash (1975) critiqued the 'golden hordes' of mass tourists causing havoc wherever they toured. However, the socio-cultural dimensions of the tourist experience were increasingly being appreciated and studied, as 'the very basis of tourism is people and society' (Sharpley, 2018, p. 4).

As Robinson and Smith (2006, p. 1) suggest, there is a complementary relationship between tourism and culture with tourism being both 'an expression and experience of culture.' Consequently, the importance of culture as a motivator for tourists and as a means of differentiating destinations became better understood, and the idea of tourism as being a cultural practice rather than a merely hedonistic experience became more commonplace. This was reflected in the 'critical turn' in academic tourism studies which began to explore a wide range of conceptual lenses through which to evaluate the practice and experience of tourism (Ateljevic, Morgan and Pritchard,

2013). Thus, the appreciation of tourism as not merely an industry, but as a socio-cultural phenomenon has led to the development of several theoretical approaches to its study which will be explored later in this chapter. Before these are examined however, it is necessary to situate the ensuing discussion within the wider context of cultural tourism, a practice which offers culturally motivated visitors the chance to experience the built heritage, places with historical association, cultural events, and the arts, all of which are also key facets of literary fiction tourism. The relationship between these touristic forms has been discussed, with some suggesting that literary tourism is a form of heritage tourism (Squire, 1994; Herbert, 2001) as it is place-based and depends on authenticity for its power. Others believe that it is more akin to cultural tourism as it depends on the creation of cultural values associated with writers (Robinson and Andersen, 2002a; Fox, 2008). Others propose that it is part of media-based tourism (Busby and Klug, 2001) with its relationships to other formats such as film (see Chapter Nine). In this study, it is acknowledged that the places and experiences that are examined are part of the historic fabric of places but also express their contemporary culture. Thus, we will examine literary tourism within the wider context of cultural tourism, using Smith's definition of this activity as 'passive, active and interactive engagement with heritage, arts, and the cultures(s) of communities' (2016, p. 17).

The evolution of cultural tourism

Chapter Two clearly demonstrates that from the Grand Tour onwards, the appropriation and appreciation of culture have long been key motivators for travel, at least for the elite for whom travel was affordable. Touring to learn languages, appreciate art, music, literature, and architecture, to visit historic monuments and even to meet important cultural icons, including authors, was the domain of the wealthy and well-educated. These cultural sites and artforms were traditionally part of what was considered to be authentic high culture, requiring taste, education, and cultural capital to be appreciated. However, in the last three decades, the relationship between culture and tourism has undergone some dramatic changes with culturally motivated travel becoming more accessible to a wider range of visitors who are able to travel to enjoy artistic and creative locations and experiences. Cultural tourism, as this form of travel has become known, is characterised by the UNWTO (2017) as 'a type of tourism activity in which the visitor's essential motivation is to learn, discover, experience and consume the tangible and intangible cultural attractions/products in a tourism destination.'

This broad definition covers a range of activities, sites, and experiences, some of which the Grand Tourist may not have considered, and demonstrates that culture is pervasive in modern tourism, so much so that the UNWTO (2018) estimates that cultural tourism accounts for almost 40% of all international tourist trips undertaken. Du Cros and McKercher (2015) have identified 12 cultural tourism product categories that include transport, creative industries (which includes literature), religious, and dark themes. This explosion of cultural tourism activity is of course linked to increased mobility and a greater demand for travel, during which culture is consumed alongside other activities (Richards, 2021). Research carried out over 30 years by the Association for Tourism and Leisure Education and Research Cultural Tourism Project (ATLAS, 2021) confirms that amongst the visitors at cultural tourism sites, approximately 10% are specifically motivated by a desire to consume culture and chose the destination for that specific purpose while the rest state culture as one of many different reasons to visit a location (Richards, 2018). Those highly motivated 'purposeful' cultural tourists (Du Cros and McKercher, 2015, p. 12) are knowledgeable about the sites visited and have a deep cultural experience whilst *in situ*, an engagement akin to that of the literary pilgrim.

The current proliferation of cultural tourism is also linked to ways in which cultural production has changed over the centuries and the expanded definition of what constitutes cultural consumption itself. Sacco, Ferilli, and Tavano Blessi (2018) describe these shifts in terms of three eras: Culture 1.0, Culture 2.0, and Culture 3.0. In the pre-industrial time of Culture 1.0, cultural products and experiences were not available to the masses but were instead a by-product of status and later of industrial growth where wealthy industrialists invested in culture to enhance their image or show benevolence to the community through the provision of public museums, galleries, and libraries. In the Culture 1.0 era, culture did not generate its own wealth but was subsidised by external funding sources. However, by the time of Culture 2.0, culture is increasingly being seen as an industry in its own right with public sector policy and investment intended to stimulate economic growth, regenerate post-industrial cities, and create jobs. Flagship developments such as the Guggenheim Museum in post-industrial Bilbao, Spain, or the Tate Modern gallery in London, housed in a disused power station and one of the UK's most popular visitor attractions, are both examples of Culture 2.0 production. The politics of the heritage industry in this era provoked much analysis (Smith, 2016) with Robert Hewison lamenting that in the mid-1980s, museums were the UK's new growth industry whilst factories and mines were in decline (Hewison, 1987).

Although the great cultural institutions and sites produced in the Culture 1.0 and Culture 2.0 eras are still very popular features in tour itineraries, a wider definition of culture in the Culture 3.0 era means that previously overlooked features of everyday life now become valued by cultural tourists. There is an emphasis on local destination culture such as street markets, community festivals, and local creative outputs where visitors may feel they are engaging in authentic experiences rather than simply gazing at the iconic sites of a traditional cultural tour (Richards, 2021). A curatorial role for cultural mediators has emerged during this era where amassed digital information can be scanned, selected, and presented to consumers, just as cultural itineraries have long been created for visitors by more traditional tourism agencies. This form of cultural curation 'becomes an act of placemaking, highlighting locations that tourists can identify with and where they can experience meaningful things' (Richards, 2021, p. 9). Tourists themselves can engage in digital content creation which is shared amongst fellow-travellers and used by destination management organisations, thus contributing to a wider sense of place, and engaging the visitor in the co-creation of the tourist experiences as we shall see later in the chapter.

In their recent report on the synergies between culture and tourism, the UNWTO (2018) also outlines the changing nature of this relationship suggesting that there has been, *inter alia*, a general shift from elite to popular culture and consequently a move away from monumental and tangible sites towards intangible forms of cultural experience. This has placed an emphasis on the culture of the everyday rather than the spectacular and a focus on the 'eventification' of cultural tourism to generate interest and demand. This involves creating a sense of occasion through highlighting specific cultural happenings or promoting exhibitions as once-in-a-lifetime exhibitions – what Hauptfleisch (2016, p. 36) refers to as 'framing the ordinary as extraordinary.' Along with this reconceptualisation of what constitutes acceptable raw material for cultural tourism, the UNWTO (2018) report also notes the increasing fragmentation of the sector into a series of identifiable niches such as heritage tourism, film and television tourism, and literary tourism. These sub-sectors, often based on forms of popular culture, reflect the increasing specialisation seen across the whole tourism industry and are a new means by which visitors can differentiate themselves and demonstrate their taste as will be seen later in this chapter. The increasingly diverse nature of the cultural tourism phenomenon is also reflected in academic research on the topic which, according to Richards (2021), has expanded from a handful of studies in the 1980s to an estimated 9000 publications a year, representing almost 10% of all tourism studies published. The evolution of cultural and heritage tourism must also be considered within the context of the United Nations' Sustainable Development Goals (United Nations, 2022) as these tourism niches tend to utilise and conserve existing infrastructure and empower communities through valuing local culture (see Chapter Ten).

It is intended that this brief overview of the development of cultural tourism has set the scene for an examination of a selection of theoretical perspectives on the study of both cultural and literary fiction tourism. These are outlined below and subsequently referred to in specific examples and cases in later chapters. In Box 3.1, the evolution of literary tourism as a subject of academic study is outlined and the key themes and theoretical frameworks which have been used in publications dating from 1997 onwards are presented. It can be seen from this case study that the concepts of pilgrimage and authenticity have long been popular lenses through which researchers have viewed literary tourism and these perspectives, along with several other theoretical frameworks are explored in more detail below, namely: the tourist gaze, tourism's experiential turn, commodification, and cultural capital.

Box 3.1 Case study: Literary tourism as a topic for study

Although we have seen in Chapter Two that literary-inspired travel has been around for thousands of years, the academic study of this phenomenon is a relatively new activity with published research appearing only from the 1980s onwards in the form of much-cited studies such as Butler (1986), Pocock (1992), Squire (1994), and Herbert (1996).

In 2020, Samet Çevik published a paper entitled 'Literary tourism as a field of research over the period 1997–2016' in the *European Journal of Tourism Research*. This paper sets out to analyse the academic output in the field of literary tourism studies and provides interesting reading. Çevik found that there had been 132 papers on literary tourism published in academic journals between 1997 and 2016. These papers were almost exclusively empirical rather than conceptual in nature, and used largely qualitative methods, in particular the case study. The geographic spread of these publications showed that research in literary tourism is predominantly a European phenomenon (with a strong bias to the UK). 70 of the 132 papers were European with the second most prolific world region being Oceania with 19 papers. Asia has 17, the Americas has 16 and, in this period, there were only five articles focussed on African literary tourism.

The typologies of literary tourism forms have been explored in this book in Chapter One (see Table 1.1). Çevik used these categories to uncover which type of literary tourism formed the focus of the research and found that the most popular literary tourism type for researchers was the lure of areas because they were appealing to literary and other figures. The next most researched aspect was literary tourism to places of significance in the work of fiction. The third most popular focus was travel-writing, then film-induced literary tourism. Interestingly, aspects of homage to an actual location (for example authors' homes, workspaces, or graves) only featured in 22 papers. By far the least popular aspects of literary tourism from a researcher's point of view were literary festivals and bookshop tourism.

The author also carried out content analysis to ascertain the research themes of the output and uncovered several approaches to the research. Authenticity was a key concept utilised to explore literary tourism, as were themes around the motivation, experience, and literary awareness of visitors. Understandably, management, marketing, and interpretation issues featured largely but there was also a focus on the concept of literary tourism and pilgrimage. Literary trails and tours as discrete products and experiences also received considerable attention within the publications.

A further scan of the published literature since 2016, using the same keyword search used in Çevik's research above, reveals that there have been in excess of 60 further academic papers in the field published in the five years between 2017–2021. Clearly interest in literary tourism has gained recent momentum. The papers show a more balanced distribution amongst the world's regions and along with the papers on the development, marketing, and management of literary tourism we also find more conceptual studies on, for example, embodiment, co-creation, touristic poetics, and contents tourism. We also find more studies exploring the role of book towns, bookshops, and literary festivals than were found in the earlier research period.

Literary tourism as pilgrimage

Tourism is frequently associated with the concept of pilgrimage, which is a practice with ancient roots and common across all religions whereby the pilgrim makes a physically challenging and spiritually transcendent journey towards a sacred end point. As Knox *et al.* (2014, p. 235) suggest, 'the metaphor of the tourist as pilgrim and tourism as a pilgrimage has been an important idea in tourism studies … and popular accounts' and this is unsurprising given that pilgrimages were the earliest forms of organised human travel and represent an important precursor to modern tourism. Certainly, the practical elements of the ancient pilgrimage – the establishment of walking routes and associated hostels and other pilgrim accommodation – clearly form the basis of contemporary recreational tourism offerings. In England, for example, there are many old public houses and hotels still bearing the name The Pilgrim which attest to this, for example the fifteenth century George and Pilgrims Hotel in Glastonbury, Somerset which was built to accommodate pilgrims to Glastonbury Abbey.

Clearly, religious pilgrims on the devotional trail are travellers and, on a practical level, will share common characteristics with more recreational tourists. However, conceptually, the idea of taking a long and difficult journey towards a sacred site to demonstrate religious devotion also translates well into the contemporary tourist experience as Turner and Turner (1978) explained with their idea of liminality, *communitas*, and the tourist experience. In this famous study, they acknowledge the presence of touristic characteristics in traditional pilgrimages and relate religious pilgrim practice to tourism behaviour: firstly, the pilgrim practice of separation from the secular to the religious sphere is likened to the traveller leaving their mundane life behind in search of the extraordinary through travel. Secondly, the crossing of the *limen* or threshold represents the in-between space where the pilgrim is released from the ties of the quotidian and is open to spiritual experiences. Thus, they join the *communitas,* the classless society of pilgrims with whom they will share transcendent experiences, liberated by the 'homogenization of status' and 'simplicity of dress and behavior' of the group (Turner and Turner, 1978, p. 34). The authors liken this process to that of a tourist joining a group tour within the institutionalised and ritualised context that is contemporary tourism. Finally, the stage of reintegration into secular life at the pilgrimage's end is likened to the return to home and everyday routines experienced by the recreational traveller. Thus 'a tourist is half a pilgrim, if a pilgrim is half a tourist' in the Turners' oft-quoted phrase (Turner and Turner, 1978, p. 20). This gives rise to the possibility of both religious and secular pilgrims, and this has been explored, for example in Nilsson (2018) and Devereux and Carnegie (2006).

Both religious and secular pilgrims are still found following the ancient pilgrim journeys of the Shikoku route in Japan, the pilgrim trail to Walsingham in Norfolk, England, and the vast European network of the Camino di Santiago. The French branch, the Camino Francés, is the most popular trail and stretches for 780km from St Jean Pied du Port in France to the tomb of St James in Santiago de Compostela in north-western Spain. Visitor studies reveal that in 2010, two thirds of pilgrims on the route were Christian and the remaining third claimed to be either atheist, New Agers, or 'unspecified' (Oviedo, de Courcier and Farias, 2014). Secular visitors may appreciate the physical challenge and the opportunity to follow an attractive route although many do admit to some sort of spiritual experience by the end of the journey. This may not be surprising as 'a dominant theme within the sociological and anthropological study of tourism is the notion that tourism represents, for the tourist, a modern sacred experience' (Sharpley, 2018, p. 187) and the pilgrim trail is simply a more intense version of spiritual voyaging. This sacralisation of tourism is a by-product of the post-secular era where the resurgence of religiosity within public life is characterised by self-reflexivity and individualised spiritual practice rather than adherence to denominational belief systems, as Nilsson and Tesfahuney (2018) suggest in their analysis of the post secular pilgrim tourist. Thus this 'new age' of spirituality can be seen in the popularity of many contemporary tourism experiences, such as those listed in Baxter's (2018) article in *The Telegraph* newspaper entitled 'The world's ten most spiritual places and how to visit them.' This list includes trips to mountains, islands, and lakes along with more traditional religious sites such as temples. Thus, 'modern society has expanded what it defines as sacred, bringing about the creation of new sites of sacrality, with travel to these sites being termed pilgrimage in its own right' as new pilgrims travel in search of spiritual ideals and experiences (Olsen and Timothy, 2006, p. 5).

This conceptualisation of pilgrimage brings it into close association with the pursuit of fame where religious devotion at the shrines of saints has been replaced with equally ardent worship of the cult of celebrity. In his study of shrines and pilgrimage in contemporary society, Margry (2008) offers a definition of pilgrimage which encompasses both the spiritual and the secular and embraces this more profane devotion. Pilgrimage is therefore travel 'to a place that is regarded as more sacred or salutary than the environment of everyday life, to seek a transcendental encounter with a specific cult object for the purpose of acquiring spiritual, emotional or physical healing or benefit' (p. 17). In Margry's study, 'cult objects' may be saints' relics or, equally, places and monuments associated with political figures, musicians, even sports people and literary figures may of course be added to this list as tourists seek out places associated with writers' lives and literary output. The literary pilgrim in pursuit of their favourite author and the transformational powers of the literary pilgrimage will be considered in more detail in forthcoming chapters.

Authenticity, aura, and the literary gaze

Part of the power of the cult object discussed above is that it represents a tangible, authentic connection to the person or spiritual ideal being sought by the pilgrim tourist. Authenticity has been a central tenet of tourism studies since Dean MacCannell first published his influential text *The Tourist: A New Theory of the Leisure Class* in 1976. MacCannell disagreed with Boorstin's (1964) view that tourists were merely in search of entertainment and instead conceptualised modern tourism as a search for genuine experiences in an increasingly disaffected world. He offered us the concept of 'staged authenticity' (MacCannell, 1973) which describes the genuine-seeming culture that is presented by host communities whilst their authentic lifestyle is kept safely 'backstage' and away from tourists' eyes. Visitors continue to seek out what they believe to be genuine experiences as the holy grail of travel and thus MacCannell sees tourism as a form of pilgrimage or quest for the authentic.

Whether this quest is ever successful has been the subject of much debate in the ensuing years as tourism grew and more parts of the world became incorporated into the industry. Questions have been posed about whether the very presence of tourists has influenced the visitor experiences of places, the hosts' own culture, the nature of the host-guest relationship, and even cultural production itself (Smith, MacLeod and Hart Robertson, 2010, p. 134). Inherent in these debates was the apprehension that tourism was a debasing force that trivialised cultures and commodified genuine relationships. Of course, such critiques of tourism assume that authenticity is an objectively verifiable characteristic of people, places, and objects and that there is a universal agreement on its measurement. Clearly this is not the case as authenticity is a relative and shifting term and subsequently discussions of its role and significance have crystallised around three distinct perspectives: objective authenticity, constructive authenticity, and existential authenticity (Cohen, 1988; Wang, 1999; Jamal and Hill, 2002).

The concept of objective authenticity assumes that genuine touristic experiences exist but are hard to find in the modern world. This notion is closely related to the nostalgia with which we might consider pre-modern society – we have lost the innocence of old, and our travels represent our search to regain this golden age in our own alienated era. This is not a new impulse, of course, as the Romantic Movement of the eighteenth century prized nature and simplicity as authentic antidotes to encroaching industrialisation and set the scene for this contemporary search for the genuine within travel. Authenticity is omnipresent in tourism advertising where 'real' and 'undiscovered' are popular descriptions of many types of travel product and experience. Of course, the concept of the verifiable and genuine is also applied to specific objects and many of these form the basis of tourist trips too – visitors only wish to see the real Mona Lisa, mask of Tutankhamen, or lock of W.B. Yeats' hair and these objects, situated as they are in museums which act as official authenticators, have the aura of genuine objects.

Walter Benjamin in his text *The Work of Art in the Age of Mechanical Reproduction* (1935) referred to aura in his critique of reproduced art objects which lose this magical quality when replicated (Benjamin, 1968). He believed that an authentic, original work of art had an aura, a mystique around it caused by the distance created between the object and the viewer – a distance which allowed the viewer to contemplate it fully. Modern reproduced works (such as films, prints of artworks or photographs) do not invite contemplation, and are consumed in a state of distraction. This, according to Benjamin, degrades the experience and the original artwork from which the reproduction may have come. We still value (and tour) original artworks, historic buildings, handwritten manuscripts, and first editions of books but our everyday lives are now more dominated by mass-produced culture. But inevitably, mass-produced culture has developed its own auratic and prized objects with museums devoted to these cultural forms, whether it be galleries of Andy Warhol prints or museums dedicated to the design and packaging of consumer goods.

Our relationship with the authentic and genuine has therefore undergone a revolution in our technologically-advanced society, to the point where we experience what cultural theorist Jean Baudrillard famously termed 'hyperreality' – a condition wherein the real and the fake are so well-integrated that it is impossible to differentiate between them (Baudrillard, 2019). Umberto Eco's essay, *Travels in Hyperreality* applies the concept to a range of American visitor attractions including Disneyworld. In his assessment of the hyper-real, he concludes that in some circumstances, the highly accomplished fake version is in fact preferable to the real. Disney's animatronic pirates and model crocodiles behave in a more satisfying manner than their real-life counterparts ever could and 'Disneyland tells us that technology can give us more reality than nature can' (Eco, 1986, p. 44). Contemporary culture is replete with copies and fakes – these are found in every site of consumption from casinos and shopping malls to cruise ships and theme parks which replicate Mediterranean villages, the Sphinx, the Roman Forum, or the Eiffel Tower. We also find examples of what Baudrillard termed a *simulacrum* – an identical copy of something that never existed. Examples of this might include representations of mythical beings such as King Arthur and his knights or Santa Claus and his elves but *simulacra* can also be made of ideas about times and places such as the Wild West or the Roaring Twenties. These ideas are based on numerous overlapping media representations rather than on an authentic reality that anyone has lived through.

If objective authenticity is hard to find in contemporary society, then an alternative perspective may be the paradigm of constructive authenticity which proposes that reality is not out there, ready to be tracked down by intrepid travellers but is, in fact, merely a constructed phenomenon. We each individually establish a sense of authenticity in our own minds, influenced by our personal perspectives, but also affected by our prevailing social, cultural, and political context. Thus, our sense of authenticity is not fixed but changes over time and space. Wang suggests that, within tourism, 'authenticity is thus a projection of tourists' own beliefs, expectations, preferences, stereotyped images and consciousness onto toured objects, particularly toured Others' (Wang, 1999, p. 355). This form of emergent authenticity explains how quickly individuals and communities can create and embrace new traditions which are then consumed as authentic by visitors. At the individual level, families can immediately create traditions around festivals such as Christmas or New Year which take on the patina of age very speedily. Similarly, on a more collective basis, Hobsbawm and Ranger's (1983) book on the *Invention of Tradition* shows us that this works at a national level too with flags, national anthems, and the British coronation ceremony being clear examples of the deliberate creation of seemingly authentic traditions which are quickly absorbed into the national culture and ultimately the tourism product. The tourism industry is also adept at constructing authenticity and directing the visitor towards 'genuine' holiday experiences. The authority accorded to tourism specialists, whether they be tour operators, travel agents, tour guides, or other mediators, gives them particular powers in creating itineraries of genuine-seeming experiences and shaping the visitor's experience of destinations and their culture.

Authenticity, therefore, is a fluid, emerging quality of places and objects rather than a fixed and verifiable entity. However, there is a final perspective, that of existential authenticity where the individual encounters the genuine first-hand through their own personal experiences. This perspective on authenticity relates to the idea discussed above regarding the loss of simplicity in everyday life. Our busy, technologically driven existence and the demands of work make it hard to be true to ourselves and retain a sense of playfulness. Thus, it is perhaps only when we are free from these constraints and on holiday that we might sense our own true selves and have genuine reactions and experiences. A vacationing life often involves a simpler routine based on relaxation and human relationships and it is perhaps whilst being tourists that we are at our most authentic, especially if we are actively participating in a destination rather than passively consuming it. Ooi (2002) believes that there will be a greater sense of existential authenticity achieved away from tourism mediators where visitors might feel that they are part of the wider community. Active, bodily engagement with culture (for example learning a local sport or craft) may also create a sense of existential authenticity as Daniel (1996) explained in relation to tourist dance performances. Likewise, holidays taken in nature, such as camping or hiking trips, may also return the visitor to their essential selves and offer freedom and simplicity.

The tourist gaze

The final concept that we will explore in relation to authenticity is that of the *tourist gaze*. British sociologist John Urry published his influential work in 1990 and this much-referenced text in now available in three editions (1990, 2002, 2011). In his books, Urry utilises the concept of the 'medical gaze' which was proposed by French theorist Michel Foucault in 1963 to describe the capacity of a physician's visual gaze in shaping perception of patients and constructing their illness through the power of sight (Foucault, 1973). Urry applies this theory to tourism, suggesting that the power of the tourist's gaze can have an equally powerful and transformative effect on the places they view. The theory of the tourist gaze is the idea that 'tourists' ways of seeing places and people and the selection of those sights is directed and organized by the tourism industry' (Smith et al., 2010, p. 174). Dean MacCannell (1976) referred to this process as 'sight sacralisation' which is the process where places are first set aside as worthy of visitation. The second stage is 'framing,' where the object or site has an official boundary placed around it, thus raising it above other locations and contributing to the authenticating process. The next stage is referred to as 'enshrinement' where a special setting or context is created to facilitate tourists' appreciation of the site. The final stage of the sacralization process is what MacCannell referred to as 'mechanical and social reproduction' where images and models of the site are produced as souvenirs and advertising materials. This stage also refers to the process whereby regions may start to name themselves after these famous attractions, for example 'George Eliot Country' in Warwickshire, UK.

This process of turning sites into sights is central to Urry's concept of the tourist gaze which proposes that there is a specific touristic lens through which places are seen and appreciated. This lens is created by the tourism industry and the media, which direct the gaze to the places deemed appropriate for touristic viewing and ignore the less palatable sights. That the tourists' primary sense is visual is, of course, no surprise as from its earliest days, touring has always been organised around sightseeing. The accoutrements of tourism are essentially visual, whether they be guidebooks, scenic viewpoints, observation carriages, or hotel terraces and balconies offering the optimum view. The visual appropriation of places is continued by the purchasing of picture postcards, the taking of photographs, and the endless circuit of these images through tourist-oriented publications and, more recently, social media platforms. Urry therefore argues that, through the tourist gaze, we are taught how to see, what to appreciate and what to edit from our visual appreciation of place.

There are two types of gaze – the romantic and the collective gaze. The romantic gaze depends on the solitary nature of the experience for its impact; for example, we have been taught that appreciating a famous work of art or the sun rising over Machu Picchu is best done alone or with a significant other. Conversely, the collective gaze tells us that the correct way to appreciate the colour and bustle of a street market or a carnival is within the midst of a crowd. Thus the tourism industry aims to create a highly selective landscape upon which the tourist gazes but of course the act of looking itself is not value-free and cannot always be directed as 'one's eyes are socially-culturally framed and there are various "ways of seeing" … people gaze upon the world through a particular filter of ideas, skills, desires and expectations, framed by social class, gender, nationality, age and education' (Urry and Larsen, 2011, p. 2).

Embodying the gaze

The wholly visual nature of the gaze has been critiqued since the first edition of *The Tourist Gaze* in 1990 with commentators suggesting that tourism is a multi-sensory experience with smell, taste, touch, and sound playing a similarly important role. In privileging sight, much of the tourist experience has been ignored and a focus on the gaze suggests that tourism is a purely passive experience with visitors obediently absorbing whatever sights have been presented to them (Perkins and Thorns, 2001). Urry countered his critics by reasserting 'the awesome dominance of the visual' (2002, p. 149) but admitting that, as we become increasingly mobile, the gaze is becoming less one-dimensional and more embodied. Thus, visitors will experience a range of sense-scapes whether they are adventure tourists, engaging all their senses and bodily skills in the activity, or gastronomic visitors, experiencing place through terroir, taste, and smell. The trend for dark restaurants (such as *blindekuh* in Zurich, Switzerland) eschews the gaze entirely and focusses the diner on every sense but the visual. Academic research into the interplay between embodiment and place now includes studies of 'soundscape, smellscape, olfactory geography, and other sensory geographic studies' (Jiang and Lu, 2020, p. 128). Becoming part of the scene in a multi-sensory tourist experience rather than simply viewing the landscape as a passive gazer may produce more existentially authentic encounters and the concept of embodiment and its associations with performance may help us to understand the experience of literary tourists as we shall see in subsequent chapters (Perkins and Thorns, 2001; Rickly, 2021). This discussion of the development of tourism from a visual to a more embodied experience now leads us onto the next theoretical approach to the study of tourism to be considered in this chapter – namely, the experiential turn.

Tourism and the experiential turn

As the world's major service industry, tourism has traditionally been concerned with the provision of customer service, whether that be a comfortable seat on an aircraft, a well-appointed hotel room, or the allocation of sufficient sun-loungers by the swimming pool. This continues to be important, but we have now moved into an era where the simple provision of good service is no longer sufficient – tourism must now be in the business of providing experiences. Pine and Gilmore recognised this from 1998 onwards in their influential work on the 'experience economy.' In their studies, the authors set out their contention that humankind has lived through four economic eras – the agrarian era where we extracted commodities from the earth, the industrial era when we manufactured goods, and subsequently the service era where heavy industry gave way to the flourishing service sector. However, Pine and Gilmore (1998) believe that we are now in the fourth economic era, the experience economy, where the main economic offering is the staging of experiences. The authors believe that 'experiences are a distinct economic offering, as different from services as services are from goods'

(p. 97) and that these experiences have two dimensions. The first of these is the level of customer participation (ranging from active to passive) and the level of connection between the customer and the experience itself. This dimension ranged from absorption to immersion with the authors contending that the most engaging and transformative experiences will be those wherein the customer is actively participating in a fully immersive event. The use of the term 'customer' is deliberate here as Pine and Gilmore have the view that individuals will only value those experiences that they have paid for. They believe that the experience economy can be applied to all service providers, for example banks and estate agents, where offering café-style settings within which customers can discuss their requirements would enhance the service experience.

The concept therefore has a wide application but is particularly apt for the hospitality and tourism industries where businesses such as the Disney corporation have been providing successful models of themed experiences since the 1950s (Bryman, 2003). Thus, the tourism and hospitality industries actively embrace the experience economy and the practice of 'experiential marketing' (Schmitt, 1999) of destinations and attractions, with businesses emphasising the encounters and memories that a trip will provide, rather than specifically promoting aspects of comfort, entertainment, education, or sociability. The emphasis on experience has been seen across the tourism and leisure industries from 'sensory hotels' (such as the 'Sensatori' hotel brand which focusses on gratifying all five of their guests' senses) to immersive hand-on exhibits in museums which for so long have been considered as places of quiet and passive knowledge acquisition. Pine and Gilmore's (1998) approach to the offering of experiences is clearly a business-orientated one with the emphasis on the profit to be made from offering customers a wide array of attractive experiences. This, however, is a rather narrow view on the power of experiences. For every carefully stage-managed and professional holiday experience that we have paid for there are also many spontaneous occurrences and encounters which we chance upon, and which are often the most memorable experiences that we take home with us.

Prahalad and Ramaswamy (2004) recognise the influence that Pine and Gilmore's concept has exerted but have proposed a second generation of experiences that are based not on stage-managed entertainment, but on the notion of 'co-creation.' In a co-creation context, it is the individual's own experience of engaging with a place or a product which creates the value for them, thus the experience is created in cooperation between visitor and site. The role of a business is therefore to create an 'experience environment' where consumers can create their own personalised experiences. An active engagement with the site and other users therefore creates an experience that 'is the sum of the psychological events a tourist goes through when contributing actively through physical and/or mental participation in activities and inter-acting with other subjects in the experience environment' (Campos *et al.*, 2018, p. 391). This type of engaged activity is expected in ludic environments such as theme parks, but co-creation can also be introduced in more traditional leisure settings such as museums and heritage sites where the emphasis has conventionally been on the appreciative tourist gaze which is designed to both preserve the site and engender the correct level of reverence for history and the aura of authentic collections. Visitors can be encouraged to participate using sensory or tactile tools (handling exhibits, sound effects, recreated smells, or food tasting), storytelling (listening and encouraging visitors to write their own tales), and personalisation, for example allowing visitors to create their own personalised route through a site (Antón, Camarero and Garrido, 2018).

In the tourism studies literature, the evolution from experience economy to co-creation reaches its current apogee within the sphere of creative tourism which is 'a practice which unites consumers and producers in the ludic construction of space' (Richards, 2011, p. 1234). Creative tourism encourages visitors to pursue creative activities and develop their own skills in activities which are related to the culture of their chosen holiday destination, whether that be cookery classes in Italy, ice carving in Canada, or flamenco lessons in Spain. Tourism, once seen as a debasing influence, robbing destinations of their culture, is now seen as a means of celebrating local crafts and providing individualised experiences for visitors. Thus, creative tourism networks have been set up in many regions, encouraging stakeholders to cooperate in providing cultural itineraries and routes to facilitate visitors' creativity. In literary tourism locations, visitors are increasingly being encouraged to interact with the sites and the narratives through costumes and activities as we shall see in Chapter Nine (see Fig. 3.1).

Figure 3.1 Bonnets and quills: Performing literary fiction tourism at Jane Austen's House, Chawton, Hampshire. Photograph by the author.

Tourism, commodification, and cultural capital

As we have seen from the preceding discussion, the commodifying power of tourism has long been a central subject of academic research in the field. The act of packaging travel, places, and culture into a commercial product to be sold and consumed raises many debates around the debasement and inauthenticity of the resulting visitor experience, not to mention the negative impacts that such processes have on host communities, traditions, and artefacts. These debates have focussed on the elitist dichotomy between the educated, free ranging traveller and the much-maligned figure of the mass tourist, consuming pre-packaged, bite sized experiences. They also cluster around the quality of the visitor experience, the impact on the culture being toured, and the nature of the host-guest relationship (MacLeod, 2006, p. 177). However, as Meethan (2001, p. 5) suggests, 'the processes of commodification, rather than being a side issue, are in fact central to the whole basis of tourism and, what is more, that tourism is one aspect of the global processes of commodification rather than a separate self-contained system.' As an important aspect of our already commercialised modern lives, it is impossible to single out tourism as the only villain within this particular piece. Thus, to understand the impact that travel has on culture, one must examine consumer culture in its entirety (Watson and Kopachevsky, 1994).

Countering the critiques of touristic commodification are the many benefits that this process can bring to hosts and visitors. If properly managed, tourism can be a protector of cultural and heritage assets, raising awareness and creating respect within both the community and the visitor along with generating much-needed funds. Packaging cultural experiences can make them more accessible to a wider audience and provides a more dependable experience (Fox, 2008) where risk is minimised to maximise enjoyment. The theme park is an ideal example of this type of controlled experience as can be seen in Chapter Six.

Clearly, as a cultural manifestation, literature is susceptible to the power of commodification and of course the production of books is a huge commercial industry. The accompanying sites and experiences that have amassed around works of fiction and their authors, described in detail in Chapter One, are a further extension of this process, giving readers more than just the novels to enjoy. As we have already seen in Chapter Two, 'we have moved from passive, informal, almost incidental, encounters with literary locale to deliberate creations of literary spaces for tourists' (Robinson and Andersen, 2002b, p. 14). Early forms of literary tourism focussed on authentic, biographical sites but this soon extended to a wider range of associative settings and, in Chapter Six, we will examine the more commodified aspects of literary tourism development.

The preceding discussion touches upon the production and consumption of culture within our contemporary consumerist society. However, to consume culture is a very different activity to assimilating other types of commodities such as mass-produced goods and media, as the consumption of culture uniquely bestows upon the consumer a prestige and value that can be translated into power in the form of 'cultural capital.' French sociologist Pierre Bourdieu coined this term in his study *Forms of Capital* (published in 1986) where he proposed that culture can become a type of capital which is not solely based on material assets, but which can nonetheless convey power to those who possess it. Bourdieu categorised the concept of cultural capital into three types – embodied, institutionalised, and objectified (Bourdieu, 2011). Embodied cultural capital is built up over time through our education and our social networks and is literally conveyed bodily through our mannerisms and the skills we have acquired. Institutionalised cultural capital is the form that can be easily recognized and measured externally, for example it is found in the academic degrees that we have attained and the positions we hold within our employment. The third form of capital, objectified capital, is literally a manifestation of the material possessions that we own, the brands we choose and the books we read. In each case, it was found that home socialisation, education, and occupation are strong influencers of the levels of cultural capital that an individual will possess.

Bourdieu also believed that the manifestation of cultural capital varies across the different social classes who each have their separate habitus – a disposition created by social and economic circumstances. In his study of French society, *Distinction: A Social Critique of the Judgement of Taste* (published in 1979), Bourdieu proposed that the ruling classes' habitus was one of effortless aesthetic appreciation, the petit bourgeoisie (middle classes) had a habitus of striving and educational ambition, and the working class operated within one characterised by popular tastes and bodily enjoyment. Across these very distinct environments, culture plays a central role in societal power relations, as classes seek to distinguish themselves through their display of capital culture – or in other words, their 'taste.' Thus the 'social order is progressively inscribed in people's minds' (Bourdieu, 1984, p. 471) through the cultural products consumed and the aesthetic choices made. The notion of conveying identity and lifestyle through cultural consumption is of course not new – in 1899 Thorstein Veblen introduced the term 'conspicuous consumption' in his book *The Theory of the Leisure Class: An Economic Study in the Evolution of Institutions.* In this study, he explored the showy displays of luxury goods being used to demonstrate wealth – something once the preserve of the ruling classes but subsequently followed by the nouveau riche, keen to demonstrate their economic credentials (Veblen, 2005). Bourdieu's concept of cultural capital however is a more nuanced and indeed inconspicuous phenomenon, based on inherent aesthetic appreciation rather than display. Therefore, 'distinction in cultural taste with respect to such seeming trivialities as preferences in clothing, leisure pastimes, music or reading matter can signify social standing' (Wang, Davis and Bian, 2006, p. 316) and certainly interest in literary-inspired tourism activities can be included in the list of 'taste' markers.

An individual's accumulation of cultural capital may go hand-in-hand with economic capital as in the case of the appreciation of live opera which is both an expensive pastime and one in which high educational attainment is required to fully understand what is going on (despite live streaming and digital downloads very effectively widening opera's more recent audiences). However, some signifiers of cultural capital are related to what would once have been markers of poverty, for example the well-educated middle classes' predilection for hiking, camping, and outdoor activities which were once the domain of the indigent. Similarly, the favoured style of pared-down home-furnishing with stripped wooden floors that currently signify good taste would once have simply displayed an inability to afford carpets. Bourdieu's popular classes' preferences are for less ascetic experiences and more comfortably upholstered living environments, thus reflecting their habitus of bodily enjoyment.

With these cultural indicators, classes struggle for cultural superiority as much within tourism choices as in wider lifestyle decisions. Greg Richards (1996) suggests that 'these factions are responsible for many of the cultural and structural features of modern tourism consumption, such as the distinction between "traveler" and "tourist" and the spatial differentiation exemplified in tourism development "off the beaten track"' (p. 266). This is seen clearly in Ahmad's (2014) research into Western tourists' choices of holiday experiences in India. This study showed that young visitors from prosperous backgrounds and with high level of educational attainment were much more likely to be travelling on a shoestring and 'roughing it' in search of authenticity. Those with lower economic and cultural capital preferred a packaged experience to more popular destinations, booked through traditional tour operators. Travellers with the back-up of economic capital have the confidence to explore and take chances with their travels, safe in the knowledge that they can afford to get themselves out of trouble if required. Consequently, experiencing the tribulations of the unbeaten track conveys social cachet and is the prerogative of the wealthy. In terms of literary-inspired tourists, distinction is conveyed through cultural choices such as one's reading materials and visits to locations associated with canonical, and more obscure authors. As we shall see in Chapter Seven, destinations can also gain cultural capital through their literary reputations which are often strenuously cultivated.

Conclusion

In this chapter we have situated our study of literary tourism within the wider context of cultural tourism, tracing its development from visits to famous monuments and institutions to a more populist appreciation of local culture. Having followed this trajectory, the chapter continued by exploring a number of theoretical approaches which have been used to inform the understanding of the practice of touring. These approaches can also usefully be applied to the study of literary fiction tourism as we shall see in subsequent chapters. The concepts of pilgrimage, authenticity, the tourist gaze, the experiential turn, commodification, and cultural capital can help us to understand the motivations, choices, and experiences of the literary tourist and to evaluate the development of literary themes within the tourist destination. Over the next five chapters we will explore the different forms of literary fiction tourism which feature in this book (see Table 1.2), commencing with Chapter Four in which we visit a range of biographical literary places – spaces intimately associated with an author's life.

References

ABTA (2020) Travel Trends Report, Association of British Travel Agents, Available at: https://www.abta.com/industry-zone/reports-and-publications/abta-travel-trends-reports/travel-trends-2020 (Accessed: 3 July 2022).

Ahmad, R. (2014) 'Habitus, capital, and patterns of taste in tourism consumption: A study of Western tourism consumers in India', *Journal of Hospitality & Tourism Research*, 38 (4), pp. 487–505. Available at: doi:10.1177/1096348012461550.

Antón, C., Camarero, C. and Garrido, M.J. (2018) 'Exploring the experience value of museum visitors as a co-creation process', *Current Issues in Tourism*, 21 (12), pp. 1406–1425. Available at: https://doi.org/10.1080/13683500.2017.1373753 (Accessed: 10 July 2022).

Ateljevic, I., Morgan, N. and Pritchard, A. (eds.) (2013) *The critical turn in tourism studies: Creating an academy of hope*. Abingdon: Routledge.

ATLAS (2021) ATLAS Special Interest Groups – Cultural Tourism Research Group. Available at: http://www.atlas-euro.org/sig_cultural.aspx (Accessed: 5 July 2022).

Baudrillard, J. (2019) 'Simulation and simulacra' (first published in 1981), in Greer, C. (ed.) *Crime and media: A reader*. Abingdon: Routledge, pp. 69–85.

Baxter, S. (2018) 'The world's 10 most spiritual places and how to visit them', *The Telegraph*, 22 February. Available at: https://www.telegraph.co.uk/travel/lists/best-spiritual-places-to-find-yourself/ (Accessed: 5 July 2023).

Benjamin, W. (1968) 'The work of art in the age of mechanical reproduction' (translated by H. Zohn), in Arendt, H. (ed.) *Illuminations*. New York: Schocken Books, pp. 217–252.

Boorstin, J.D. (1964) *The image: A guide to pseudo-events in America*. New York: Harper and Row.

Bourdieu, P. (1984) *Distinction: A social critique of the judgement of taste*. Translated by R. Nice. Harvard: Harvard University Press.

Bourdieu, P. (2011) 'The forms of capital' (first published in 1986), in Szeman, I. and Kaposy, T. (eds.) *Cultural theory: An anthology*. Hoboken, NJ: Wiley, pp. 81–93.

Bryman, A. (2003) *Disney and his worlds*. Abingdon: Routledge.

Busby, G. and Klug, J. (2001) 'Movie-induced tourism: The challenge of measurement and other issues', *Journal of Vacation Marketing*, 7 (4), pp. 316–332.

Butler, R.W. (1986) 'Literature as an influence in shaping the image of tourist destinations: A review and case study', in Marsh, J. (ed.) *Canadian studies of parks, recreation, and foreign lands*. Peterborough: Trent University, pp. 111–132.

Campos, A.C., Mendes, J., Valle, P.O.D. and Scott, N. (2018) 'Co-creation of tourist experiences: A literature review', *Current Issues in Tourism*, 21 (4), pp. 369–400. Available at: https://doi.org/10.1080/13683500.2015.1081158 (Accessed: 20 July 2022).

Çevik, S. (2020) 'Literary tourism as a field of research over the period 1997–2016', *European Journal of Tourism Research*, 24, pp. 1–25. Available at: https://doi.org/10.54055/ejtr.v24i.409 (Accessed: 28 July 2022).

Cohen, C. (1988) 'Authenticity and commoditisation in tourism', *Annals of Tourism Research*, 15 (3), pp. 371–386. Available at: https://doi.org/10.1016/0160-7383(88)90028-X (Accessed: 20 July 2022).

Daniel, Y.P. (1996) 'Tourism dance performances authenticity and creativity', *Annals of Tourism Research*, 23 (4), pp. 780–797. Available at: https://doi.org/10.1016/0160-7383(96)00020–00025 (Accessed: 20 July 2022).

Devereux, C. and Carnegie, E. (2006) 'Pilgrimage: Journeying beyond self', *Tourism Recreation Research*, 31 (1), pp. 47–56. Available at: https://doi.org/10.1080/02508281.2006.11081246 (Accessed: 4 February 2022).

Du Cros, H. and McKercher, B. (2015) *Cultural tourism*. 2nd edn. Abingdon: Routledge.

Eco, U. (1986) *Travels in hyper-reality*. London: Picador.

Foucault, M. (1973) *The birth of the clinic*. London: Tavistock Publications.

Fox, R. (2008) 'Marin Držić: A case for Croatian literary tourism', *Journal of Heritage Tourism*, 3 (2), pp. 138–153. Available at: doi:10.1080/17438730802138154.

Hauptfleisch, T. (2016) 'Eventification: Framing the ordinary as the extraordinary', in Pernecky, T. (ed.) *Approaches and methods in event studies*. London: Taylor & Francis, pp. 36–52.

Herbert, D.T. (1996) 'Artistic and literary places in France as tourist attractions', *Tourism Management*, 17 (2), pp. 77–85. Available at: https://doi.org/10.1016/0261-5177(95)00110–00117 (Accessed: 3 November 2021).

Herbert, D.T. (2001) 'Literary places, tourism and the heritage experience', *Annals of Tourism Research*, 28 (2), pp. 312–333. Available at: https://doi.org/10.1016/S0160-7383(00)00048–00047 (Accessed: 3 November 2021).

Hewison, R. (1987) *The heritage industry: Britain in a climate of decline*. London: Methuen.

Hobsbawm, E. and Ranger, T. (eds.) (1983) *The invention of tradition*. Cambridge: Cambridge University Press.

Jamal, T. and Hill, S. (2002) 'The home and the world: (Post) touristic spaces of (in) authenticity?', in Dann, G. (ed.) *The tourist as a metaphor of the social world*. Wallingford: CABI, pp. 77–107.

Jiang, L. and Yu, L. (2019) 'Consumption of a literary tourism place: A perspective of embodiment', *Tourism Geographies*. 22 (1), pp. 127–150. Available at: doi:10.1080/14616688.2019.1586985.

Knox, D., Hannam, K., Olsen, D.H. and Salazar, N.B. (2014) 'Is the tourist a secular pilgrim or a hedonist in search of pleasure?', *Tourism Recreation Research*, 39 (2), pp. 235–267. Available at: doi:10.1080/02508281.2014.11081769.

MacCannell, D. (1973) 'Staged authenticity: Arrangements of social space in tourist settings', *American Journal of Sociology*, 79 (3), pp. 589–603.

MacCannell, D. (1976) *The tourist: A new theory of the leisure class*. Berkeley, CA: University of California Press.

MacLeod, N. (2006) 'Cultural tourism: Aspects of authenticity and commodification', in Smith, M. and Robinson, M. (eds.) *Cultural tourism in a changing world*. Bristol: Channel View Publications, pp. 177–119.

Margry, P.J. (ed.) (2008) 'Secular pilgrimage: A contradiction in terms?', in Margry, P.J. (ed.) *Shrines and pilgrimage in the modern world: New itineraries into the sacred.* Amsterdam: Amsterdam University Press, pp. 13–46.

Meethan, K. (2001) *Tourism in global society: Place, culture, consumption.* Basingstoke: Palgrave.

Nilsson, M. (2018) 'Wanderers in the shadow of the sacred myth: Pilgrims in the 21st century', *Social & Cultural Geography*, 19 (1), pp. 21–38. Available at: https://doi.org/10.1080/14649365.2016.1249398 (Accessed: 10 December 2021).

Nilsson, M. and Tesfahuney, M. (2018) 'The post-secular tourist: Re-thinking pilgrimage tourism', *Tourist Studies*, 18 (2), pp. 159–176. Available at: https://doi.org/10.1177/1468797617723467 (Accessed: 11 December 2021).

Olsen, D.H. and Timothy, D.J. (2006) 'Tourism and religious journeys', in Olsen, D.H. and Timothy, D.J. (eds.) *Tourism, religion and spiritual journeys.* London: Taylor & Francis, pp. 17–38.

Ooi, C.S. (2002) *Cultural tourism and tourism cultures: The business of mediating experiences in Copenhagen and Singapore.* Copenhagen: Copenhagen Business School Press.

Oviedo, L., De Courcier, S. and Farias, M. (2014) 'Rise of pilgrims on the Camino to Santiago: Sign of change or religious revival?', *Review of Religious Research*, 56 (3), pp. 433–442. Available at: https://doi.org/10.1007/s13644-013-0131-4 (Accessed: 11 December 2021).

Perkins, H.C. and Thorns, D.C. (2001) 'Gazing or performing? Reflections on Urry's tourist gaze in the context of contemporary experience in the antipodes', *International Sociology*, 16 (2), pp. 185–204.

Pine, B.J. and Gilmore, J.H. (1998) 'Welcome to the experience economy', *Harvard Business Review*, 76 (4), pp. 97–105.

Pocock, D. (1992) 'Catherine Cookson country: Tourist expectation and experience', *Geography*, 77 (3), pp. 236–243. Available at: https://www.jstor.org/stable/40572194 (Accessed: 3 April 2021).

Prahalad, C.K. and Ramaswamy, V. (2004) 'Co-creating unique value with customers', *Strategy & Leadership*, 32 (3), pp. 4–9.

Richards, G. (1996) 'Production and consumption of European cultural tourism', *Annals of Tourism Research*, 23 (2), pp. 261–283. Available at: https://doi.org/10.1016/0160-7383(95)00063-00061 (Accessed: 2 December 2021).

Richards, G. (2011) 'Creativity and tourism: The state of the art', *Annals of Tourism Research*, 38 (4), pp. 1225–1253. Available at: https://doi.org/10.1016/j.annals.2011.07.008 (Accessed: 2 December 2021).

Richards, G. (2018) 'Cultural tourists: Profiles, motivations and activities', *Proceedings of the third cultural heritage seminar: Tourism and cultural heritage confluences*, Barcelona, 8 November 2018. (Vol. 8). Available at: https://www.researchgate.net/publication/328829336_Cultural_tourists_Profiles_Motivations_and_Activities (Accessed: 2 July 2022).

Richards, G. (2021) *Rethinking cultural tourism.* Cheltenham: Edward Elgar Publishing.

Rickly, J.M. (2021) 'Performing beyond the view: Embodiment and tourist experiences', in Sharpley, R. (ed.) *Routledge handbook of the tourist experience.* Abingdon: Routledge, pp. 140–152.

Robinson, M. and Andersen, H.C. (2002a) 'Introduction', in Robinson, M. and Andersen, H.C. (eds.) *Literature and tourism: Reading and writing tourism texts.* London: Continuum, pp. xiii–xix.

Robinson, M. and Andersen, H.C. (2002b) 'Reading between the lines: Literature and the creation of touristic spaces' in Robinson, M. and Andersen, H.C. (eds.) *Literature and tourism: Reading and writing tourism texts.* London: Continuum, pp. 1–38.

Robinson, M. and Smith, M. (2006) 'Politics, power, and play: The shifting contexts of cultural tourism', in Smith, M. and Robinson, M. (eds.) *Cultural tourism in a changing world.* Bristol: Channel View Publications, pp. 1–18.

Sacco, P.L., Ferilli, G. and Tavano Blessi, G. (2018) 'From culture 1.0 to culture 3.0: Three socio-technical regimes of social and economic value creation through culture, and their impact on European Cohesion Policies', *Sustainability*, 10 (11), pp. 1–23. Available at: https://doi.org/10.3390/su10113923 (Accessed: 15 November 2021).

Schmitt, B. (1999) 'Experiential marketing', *Journal of Marketing Management*, 15 (1–3), pp. 53–67. Available at: https://doi.org/10.1362/026725799784870496 (Accessed: 3 November 2021).

Sharpley, R. (2018) *Tourism, tourists, and society.* Abingdon: Routledge.

Smith, M.K. (2016) *Issues in cultural tourism studies.* 3rd edn. Abingdon: Routledge.

Smith, M., MacLeod, N. and Hart Robertson, M. (2010) *Key concepts in tourist studies* London: Sage.

Squire, S.J. (1994) 'The cultural values of literary tourism', *Annals of Tourism Research*, 21 (1), pp. 103–120. Available at: https://doi.org/10.1016/0160-7383(94)90007-90008 (Accessed: 3 February 2023).

Turner, L. and Ash, J. (1975) *The golden hordes: International tourism and the pleasure periphery.* London: Constable.

Turner, V. and Turner, E. (1978) *Image and pilgrimage in Christian culture: Anthropological perspectives.* New York: Columbia University Press.

United Nations (2022) Make the SDGs a reality. Available at: https://sdgs.un.org/goals (Accessed: 13 September 2022).

UNWTO (2017) Tourism and culture. Available at: https://www.unwto.org/tourism-and-culture (Accessed: 5 July 2023).

UNWTO (2018) Tourism and culture synergies. Madrid: UNWTO. Available at: https://doi.org/10.18111/9789284418978 (Accessed: 12 January 2022).

UNWTO (2022) Glossary of tourism terms. Available at: https://www.unwto.org/glossary-tourism-terms (Accessed: 12 January 2022).

Urry, J. (1990) *The tourist gaze 1.0.* London: Sage.

Urry, J. (2002) *The tourist gaze 2.0.* London: Sage.

Urry, J. and Larsen, J. (2011) *The tourist gaze 3.0.* London: Sage.

Veblen, T. (2005) *The theory of the leisure class: An economic study of institutions.* First published in 1899. London: Penguin.

Wang, N. (1999) 'Rethinking authenticity in tourism experience', *Annals of Tourism Research*, 26 (2), pp. 349–370. Available at: https://doi.org/10.1016/S0160-7383(98)00103–0 (Accessed: 6 November 2021).

Wang, S., Davis, D. and Bian, Y. (2006) 'The uneven distribution of cultural capital: Book reading in urban China', *Modern China*, 32 (3), pp. 315–348. Available at: doi:10.1177/0097700406288178.

Watson, G.L. and Kopachevsky, J.P. (1994) 'Interpretations of tourism as commodity', *Annals of Tourism Research*, 21 (3), pp. 643–660. Available at: https://doi.org/10.1016/0160-7383(94)90125–90122 (Accessed: 4 July 2022).

4 Biographical literary fiction tourism

From cradle to grave

In the preceding chapters, the wide range of literary-inspired sites that attract visitors has been introduced and the differentiation between biographical, imaginative, socially-constructed spaces, and places of literary repute has been established. Perhaps the most powerful of these types of literary space is the biographical site – a place closely associated with the birth, life, and ultimately the death of a literary artist. In this chapter, we explore the power of birthplaces, homes, haunts, workspaces, and graves as repositories of biographical meaning for readers and visitors. Personal belongings and relics are also considered as extensions of these locations. In considering the power of these sites associated with an author's life, authenticity is a key feature underpinning the development, presentation, and experience of these settings. The concept of pilgrimage will also be employed to further understand the lure of places where writers lived, worked, and died.

Birthplaces

The site of a famous author's birth is one key element in the array of biographical settings available for literary tourists to enjoy. These sites of origin are of course as diverse as the authors who entered the world within them and tourists seeking inspiration from these native places can find it in Leo Tolstoy's grand birthplace at the family's country estate near Tula in Russia, or in the humble cottage in the backstreets of Odense in Denmark where Hans Christian Andersen first drew breath. Ernest Hemingway was born in a comfortable Victorian mansion in Chicago, Illinois which provides contrast to the unmarked second floor flat that was Jack Kerouac's birthplace in Lowell, Massachusetts or the patch of field and water pump that marks the site of Steventon Rectory in Hampshire, the place of Jane Austen's entry into the world.

When the term 'birthplace' is used in relation to a literary theme it will, inevitably, invoke what is arguably the most famous site of authorial nativity, Shakespeare's birthplace in Stratford-upon-Avon. In this chapter we will evaluate the role of this and other significant birthplace sites in creating the idea that the place where an author first entered the world is worth noting

DOI: 10.4324/9781003041740-4

and perhaps preserving for visitors' interest and entertainment. In his article on 'The Birth of the Birthplace,' Santesso (2004) compares the Stratford birthplace with the earlier site of Bread Street in London, the place of poet John Milton's birth. He uses both these places of literary pilgrimage to illustrate the two historic models of tourism described in Chapter Two, namely the early discourse-centred form of travel where visitors sought out living artists and thinkers and the modern practice of visual appropriation through sightseeing. The Bread Street site, now long demolished, had nothing with which to divert the would-be literary tourist other than its association with the birth of the great man in 1608 and early accounts tell of visitors coming to see the poet and the chamber in which he was born (Santesso, 2004, p. 383). The Stratford model, on the other hand, provided a range of attractions and events to amplify the experience of the birthplace and augment the gaze as we shall see later in the chapter.

In her detailed analysis of the evolution of the literary birthplace, Nicola Watson (2006) suggests that these preserved shrines are less commonplace than writers' houses because the site of birth may have been developed, repurposed, or demolished before its significance as the 'cradle of literary genius' was realised. There are some notable birthplaces which have been preserved and presented for visitors as will be seen later in this chapter, but the place of birth is a less celebrated phenomenon than the adult home which housed, nurtured, and inspired the literary genius and their output. It is worth noting, for example, that amongst the 215 blue plaques in London celebrating places associated with literary artists (see Box 1.1), only a handful indicate an author's birthplace. The births of Robert Graves and Georgette Heyer (both coincidentally having occurred in Wimbledon, Southwest London) are celebrated with plaques, as is the house in Dulwich in Southeast London where Gerald Durrell spent his childhood days. Perhaps these plaques, attached to domestic dwellings on quiet streets, provide only passing interest to visitors although some may stop and ponder on the contrast between the author's first houses and their chosen domains of later life. This might be particularly marked in the cases of both Graves and Durrell who swapped the grey streets of suburban London for colourful and well-documented lives on the islands of Mallorca and Corfu.

The sites where authors happened to have been born, or indeed where they chanced upon death are demonstrably different to the homes where authors have chosen to live and work in later lives. The birthplace and the grave mark both ends of the author's life and visiting these sites invokes a sense of pilgrimage – the birthplace visit recalls the Christian pilgrimage to Bethlehem and communing at the grave models the Catholic tradition of the veneration of relics (Watson, 2006, p. 57). But whereas the grave offers physical evidence of the writer's existence, presented within confines that are demarcated and named and where the accepted behaviour is known – whether that be silent contemplation or the placing of flowers or other offerings – Hoppen, Brown and Fyall (2014), suggest that to create a memorial out of a birthplace 'requires much more effort of one's imagination, which is why the extension of emotional investment from an author's grave to his or her birthplace only manifests itself for relatively few authors' (p. 41). To denote the authorial place of birth is to place all the significance upon the environment and its future influence on the shaping of the literary genius. In some birthplaces, such as Thomas Hardy's cottage in Upper Bockhampton in Dorset, England this is much easier to grasp than in many other birth sites. Hardy was born in the cottage in 1840 and lived there for many years, inspired by the landscape around him and the views from his bedroom window. The cottage is managed by the National Trust who suggest that 'here, at a desk by the window from which he was inspired by the scenery outside, Thomas wrote his early novels including Under the Greenwood Tree and Far from the Madding Crowd' (National Trust, 2022a). Similarly, in Shaoxing, Zhejiang Province stands one of China's earliest literary tourism sites, Lu Xun's Native Place, which opened in 1953 in (see Box 2.1). The site, which covers 14,000 square metres, contains the home where Lu Xun, famous novelist, and essayist, was born and spent the years of his childhood and teenage. The area also contains his school, the homes of his wider family, and the Baicao garden which featured in his writing (Wang and Zhang, 2017). Thus, the literary pilgrim can view the full extent of his upbringing and influences from birth to early adulthood and beyond.

For many other sites, however, the idea of the authorial birthplace is more fraught. At their most basic level, problems with the concept of the birthplace centre around the fact that the actual location of the birth may be unknown or contested and the actual dwelling may have been very humble, impermanent, and unsuitable for significant development. In the case of Sam Clements' (Mark Twain's) birthplace in Florida, Missouri, no-one is sure if the rudimentary 1830s

Figure 4.1 Lu Xun's Native Place, Shaoxing, Zhejiang Province, China. Photograph by A.M. Morrison.

two-room dwelling on display within a purpose-built museum is his actual birthplace (Lowe, 2012) but the building preserved under cover is certainly representative. Nathaniel Hawthorne's authentic Georgian birthplace house was also displaced, in this case, to join other buildings associated with his life in Salem, Massachusetts. The house was moved in its entirety to the nearby House of the Seven Gables Museum campus in 1958 (House of the Seven Gables, 2022), so that the house of his birth is authentic, but the actual geographical co-ordinates of the event can be found further down the street. It is also the case that very few, if any, artefacts may have survived to animate the birthplace and provide context for visitors as their significance will not have become clear until later in the author's life. However, as we shall now discover, in the early cases of both William Shakespeare in England and Robert Burns in Scotland, much effort has gone into preserving and celebrating the birthplaces, largely down to their identity with a national culture (Watson, 2006).

Shakespeare and Burns – the birth of the national poets

Shakespeare's birthplace in Henley Street in Stratford-upon-Avon is the earliest example of an authorial native site in the United Kingdom (UK) and was a shrine for visitors long before the Shakespeare Birthplace Trust purchased the building in 1847 as a national memorial. The house had been in the ownership of four consecutive householders since the death of the playwright's sister Judith (Young, 2019) and these owners were happy to exploit the connections and dispose gainfully of the fixtures and fittings, for example a chair on which the Bard may have sat and which 'displayed a miraculous ability to regenerate itself' (Rosenthal, 2008, p. 34). Latterly, souvenirs carved from the wood of a mulberry tree that William was purported to have planted were many and various and included toothpicks, snuff boxes, and spectacle cases (Ousby, 1990, pp. 39–42). Literary tourists on the trail of Shakespeare would also visit his grave in Holy Trinity Church, Stratford (where he was interred in 1616) or his monument in Poets' Corner in Westminster Abbey, which was installed in 1740. However, it was the Shakespearean actor David Garrick's 'Shakespeare Jubilee' held in 1769, belatedly to celebrate the Bard's 200th birthday, that effectively connected the concept of a birthplace and the grave in an event that put Stratford on the map (Watson, 2006, p. 56) and forever connected the playwright with his hometown, the rural countryside of Warwickshire and, by extension, the national culture.

Garrick reputedly assigned a bedroom in the house on Henley Street which was known to have belonged to Shakespeare's father as the actual birthplace, festooned the window in illuminated drapes and incorporated the site into his Jubilee procession and celebratory events (Ousby, 1990). This event and the tourism that it consequently engendered has been referred to as 'textless tourism' (Watson, 2006, p. 59) as the emphasis is firmly upon the biographical sites and the fame of the national poet rather than the actual works themselves, which didn't feature in the Jubilee celebrations. Since the formation of the Birthplace Trust in 1847, the house, half-timbered in Shakespeare's day and since then faced with red brick, has been restored to its original appearance and thus 'the place we visit is an imaginative reconstruction of an historical building in which William Shakespeare may not have been born' (Rosenthal, 2008, p. 36). The house has been accepted as the birthplace of the Bard for so long that perhaps it may not matter whether we are looking at a site of objective or constructed authenticity.

Contemporary visitors to the birthplace experience an exhibition providing context for the Tudor home before entering the dwelling which is set up as a house rather than as a museum. The birth room itself is furnished with a parental bed and cradle of appropriate style but is otherwise 'an empty silent space, potential, secretive and blank, signifying the space or time before "Shakespeare," before there was anything to remember' (Watson, 2006, p. 60). Although the birthplace is the site that has always been most revered by visitors, Stratford offers other Shakespeare locations including Hall's Croft (home to William's daughter), Shakespeare's New Place (where he lived in later life), Anne Hathaway's cottage, and Mary Arden's Farm (the childhood home of Shakespeare's mother). A reading room, educational facilities, cafés, and gift shops complete the offering and for any visitor who may wish to engage with the actual works themselves, the Royal Shakespeare Theatre can be found by the river.

The significance of the authorial birthplace is reflected in works of literature where writers ponder on what, if anything is to be gained from such a literary pilgrimage. In Henry James' short story *The Birthplace* published in 1903, the author appears to be satirising the practice of literary tourism, although the tale also explores the difficult relationship between the author and the text and the role of literary biography (Zemgulys, 2008). In the story, we see the inner struggles of Morris Gedge, one-time school master who has taken on the job of caretaker of the birthplace of a famous poet who is never named but clues suggest is Shakespeare. Having tried to immerse himself in the works of the great man and in attempting to learn the facts which associate him with the house, Gedge realises that he cannot wholly embrace the site as the authentic birthplace. He takes to wandering the house after dark, feeling closer to the poet when he cannot see the embarrassing and anecdotal displays set up for visitors and thus demonstrating a tendency towards the discourse-centred approach to engaging with the literary themes rather than the mere act of sightseeing which is what he feels is offered at the birthplace. His tour guiding style becomes taciturn and minimalist until a reprimand from the Chair of the Committee, and an encounter with a young couple who understand his predicament, lead him to change his approach. By the end of the story, Gedge has resorted to imaginative narration for the willing tourists who greatly appreciate the fictional interpretation that he skilfully supplies. Whether these new tales will in time become part of the accepted history of the poet is one of the problems that Gedge struggles with by the end of the story (Nakamura, 1990). Tellingly, in his tour he describes the house as the nurturing environment for the nascent poetic talent thus: 'It is in this old chimney corner … that we see the inconceivable child gazing into the blaze of the old oaken logs and making out there pictures and stories … or poring over some scrap of an ancient ballad' (James, 2013, pp. 195–196). This conflation of the physical setting with imagined scenes demonstrates the important link between domestic place and early literary inspiration within the concept of the birthplace and it is a trope still used in the presentation of writers' homes.

The value of visiting Shakespeare's birthplace is also famously considered in Charles Dickens' novel *Nicholas Nickleby* (1839) where Nicholas' sister Kate's employer, the rather delicate Mrs Wititterly declares:

> I find I take so much more interest in his plays, after having been to that dear little dull house he was born in! I don't know how it is, but after you've seen the place and written your name in the little book … it kindles up quite a fire within one.

Her husband, less inspired, has come to the same conclusion as Gedge: 'There is nothing in the place, my dear – nothing, nothing' (Dickens, 1999, p. 340). The Shakespearean tourism product was also firmly-enough embedded in popular culture to feature in children's fiction, for example in Richmal Compton's *Just William* series. In *William and the Lost Tourist*, (written in 1926) our hero helps an attractive American tourist, Sadie Burford, who has lost her way *en route* to 'that Shakespeare guy's place' (Compton, 2016, p. 96). In pretending that his village is Stratford-upon-Avon and, warming to the deception that he is a descendant of the Bard, William helps Sadie to tick off the Shakespearean sites on her tour itinerary without the need of an actual visit. This scene neatly ties up the theatrical nature of many tourism offerings with the power of existential authenticity for visitors who are willing to believe.

Nicola Watson (2006) recounts the relative speed with which the birthplace of that other national poet, Robert Burns, was established, and suggests that the Stratford birthplace provided a template that facilitated the process. Robert Burns' birthplace cottage was another very early entrant into the field, having been a lure for visitors since his early death in 1796. The thatched stone longhouse which is situated on a road outside Ayr in Southwest Scotland became a pub soon after his death which made it easy for visitors to pay their respects. Unusually, the bed in which Burns was born is still *in situ* thanks to it being a recessed box bed and hence immovable. The birth room is presented today as a domestic scene with a cradle and spinning wheel set up by the fire. The cottage was bought by the town council of Alloway in 1818 to allow the pilgrim site to be managed as an attraction (Young, 2015) as his fame continued to spread. Burns, now considered to be Scotland's national poet and a cultural and touristic icon (Pearce, Morrison and Moscardo, 2003) wrote in both English and Scots and his poems and songs which champion the ordinary man and woman and celebrate the local landscape gave him the reputation as patriotic upholder of Scottish culture. So popular was this

Romantic figure, who died at the age of 37, that there was a sign proclaiming his place of birth erected outside the cottage by 1805 and by 1838 a visitors' album had been introduced showing a total of 350 visitors in the first month (Watson, 2006, p. 69). Burns memorabilia started to appear by the eighteenth century with W & A Smith of Mauchline producing snuff boxes, knitting accessories and pop guns to cater for all the family (Mackay and Pittock, 2011, p. 150).

The range of Burns-related sites has expanded considerably since the establishment of the birthplace and now includes a new purpose-built museum at the Burns Cottage site, the Robert Burns Centre in Dumfries, the Robert Burns House Museum in Mauchline, the Robert Burns House in Dumfries, the Batchelors' Club in Tarbolton, Ellisland Farm in Dumfries, and monuments to Burns in Kilmarnock and Edinburgh. This Burns landscape is heavily biographical but also embraces fictionalised settings such as the ruins of Alloway Auld Kirk and the Brig o' Doon, which feature in his poem *Tam O' Shanter* (1790). True pilgrims can also visit the statue of Burns' 'Highland Mary', Mary Campbell, where she stands on Castlehill in Dunoon on the Cowal peninsula, looking down the Firth of Clyde to her poet in Ayrshire. Many of these sites have been pulled together into the Robert Burns Heritage Trail which is a driving route through Dumfries and Galloway. The Burns theme extends into a range of contemporary souvenirs, and children are catered to in the 'Scots Wa-Hey' play area at the birthplace, complete with Tam O' Shanter zip wire and a miniature Burns Cottage. Scots at home, and most certainly those in exile overseas, celebrate Burns' birthday each year on 25 January where the traditions of music, speeches, poetry reading, and the addressing and subsequent consumption of the haggis keep the Bard's name well and truly alive in countless Burns Suppers across the globe (See Box 9.2). Given this wealth of Burns-related activity, it is perhaps no surprise to learn that the national poet has been confirmed in a recent report to have a value to the Scottish economy of £203m annually with an additional 'enduring brand value' of around £139.5m (Pittock and Ambroisine, 2019).

The writer's house

Although the places of birth discussed above have, of course, been situated in houses, the writer's house *per se* is a very different proposition. Here we will analyse the houses where authors of literary fiction spent some, if not all, of their adult lives. These may be houses which they built, renovated, or furnished themselves and are spaces where they wrote the works that made them famous. These are what Robinson and Andersen refer to as 'arguably the most powerful tourism resource' amongst the other literary-themed offerings available (2002, p. 15). Intensely biographical, the domestic setting within which an author lived, created their works, and welcomed guests is a very personal space which allows the visitor to feel closer to the artist and perhaps to the source of their genius. Harald Hendrix (2008) suggests that these sites can be seen as 'self-fashioning' and autobiographical in that in some cases, such as the Gothic homes built by Horace Walpole and Sir Walter Scott as a direct concretisation of their literary works, the author uses the home to project a desired public image and statement about their literary intent. However not all writers' homes can be categorised in this manner as Hendrix also examines the authorial house as direct inspiration for the works themselves. We see this in the presentation of many houses where we are led to the view from the window which inspired a setting or a theme. In Herman Melville's home, Arrowhead in Massachusetts, for example, visitors are shown the humpbacked shape of Mount Greylock, visible from his study, which is said to have given him the idea to write a story about a whale (Trubek, 2011).

Writers' homes may be rooms, apartments, terraced houses, cottages, or stately homes where an author has spent some part of their life, and which illustrate the author's levels of success at that time in their career. Some authors such as Nathanial Hawthorne lived in several homes throughout their lives, ranging from the modest to the grand and demonstrating key stages in their authorial progress. The idea of preserving and displaying a domestic space associated with a writer goes back as far as the 1540s with the putative home of Petrarch near Padua in Italy attracting visitors who were also in search of his tomb. Other early writers' houses include those of Goethe in Weimar, Germany which opened in 1885, Thomas Carlyle's house in London (1895), Henry Longfellow's Massachusetts home in 1913, and the Brontë parsonage in Haworth, Yorkshire in 1928 (see also Table 2.1). As we discovered in Chapter Two, some writers' homes were places of pilgrimage during the lifetime of the author with Goethe and Voltaire entertaining Grand Tourists in their homes in Frankfurt and Geneva. Charlotte Brontë was rather bemused by the visitors who began making their way to Haworth in the 1850s (Barnard, 2002) and William Wordsworth was initially delighted to see that his poetry had encouraged people to knock at his door, only to be somewhat overwhelmed by the consequent deluge of tourists (Donaldson, 2013). We have already learned that William Howitt's Homes and Haunts guides (1875) further encouraged the notion of visiting a writer's domestic sphere and other associated locations as a way of further understanding the writer and their output. These publications outlined the detailed visits that Howitt and his wife Mary made to living authors' homes (which could be vicariously enjoyed by their readership) and, increasingly, the homes of deceased writers which could be visited in person with impunity. It is therefore only after the removal of the author (usually through death) that the writer's house as visitor attraction comes into its own.

Watson (2020) comments on the writer's house concept as a 'transnational form native to Britain, Europe and North America' (p. 6) and, indeed, today there are examples of authorial homes across the globe. We have already seen that there are many writers' houses that date back to the eighteenth and nineteenth centuries, but the real period of growth has been much more recent with, for example, a quarter of France's 201 officially registered writers' houses have been established in the 1980s and 90s (Engelking, 2018). Similarly in Italy, 40% of their 60 writers' houses were established since 1990 (Hendrix, 2008). Mirroring this recent flowering, associations have been set up to further the cause of literary houses, the most active being France's National Federation of Writers' Houses and Literary Heritage which was established in 1997 and has over 200 properties and sites in its membership. In Hungary, the Association of Hungarian Literary Memorial Places was set up in 2008 with a network of over 60 members, predominantly writers' house museums (MIRE, 2022).

The impetus behind the establishment of writers' houses varies significantly from author to author. Some houses such as Horace Walpole's Strawberry Hill and Alexander Dumas' Château de Monte-Cristo were deliberately created by their owners as the physical expression of their Romantic and Gothic literary style and intended to be places of pilgrimage for their devotees. Other literary houses have been posthumously saved for posterity due to the intervention of the writer's families or of Trusts set up to acquire properties. Others may have been sold off by family members who inherited them only to be rescued by local authorities keen to preserve home-grown literary credentials, as was the case with the house in Illiers in Centre-Val de Loire, France, where Marcel Proust spent his childhood holidays with his aunt and uncle. The family sold the house in 1912 but it was acquired by the city and passed onto the Society of Friends of Marcel Proust and Friends of Combray to preserve and interpret as a writer's house (Sabre, 2017). A smaller number of sites are reconstructions, based on the original home of the author with authenticity being lent using original artefacts and input from the author's family or associates. Examples of this type of site include the reconstructed childhood home of Gabriel Garcia Márquez in Aracataca, Columbia, the original, his Grandparents' home where he spent his first eight years, having been destroyed in a fire. Similarly, the home of Necati Cumalı in Urla, Turkey was rebuilt in 2001 as the actual house was demolished in 1985 to make way for a road (Çevik, 2018).

Given the variety of modes of conception, it follows that writers' houses can be owned and managed in a variety of different ways. In the United States of America (USA), several houses are operated by the National Parks Service, including those belonging to L.M. Alcott, Nathaniel Hawthorne, and Edgar Allen Poe. Historical societies, local authorities, and private foundations are also involved (Trubek, 2011). In the UK, the National Trust (Europe's largest conservation charity) takes care of the homes of Agatha Christie, Thomas Hardy, George Bernard Shaw, and Virginia Woolf, amongst others. Literary societies such as the Brontë Society in Haworth, Yorkshire and the Katherine Mansfield Birthplace Society in Wellington, New Zealand are charities founded to take care of the writers' homes. Others are publicly funded museums, such as D.H. Lawrence's birthplace museum in Eastwood, or part of the city's museums service but secured on behalf of the public by individual benefactors as was the case with Victor Hugo's house in Paris which was bought by the city with the help of Paul Meurice, Hugo's novelist friend.

Linda Young suggests that there are three stages to the creation of a writers' house: first, there must be an available house; secondly, there must be funds and those prepared to champion the project; thirdly, the house must have ongoing commitment to its governance and operation (Young, 2015). These three stages go some way to explaining why some authors have a number of houses dedicated to them, others just a blue plaque affixed to a private dwelling and some none at all. The allocation of house museums to authors is 'illogic' (Trubek, 2011, p. 6) and certainly not correlated with a writer's fame or output. In the UK, the much beloved and prolific Anthony Trollope has no museum and Elizabeth Gaskell's home in Manchester has only been a visitor attraction since 2014. In the USA, literary greats such as J. D. Salinger, Saul Bellow, Kurt Vonnegut, and even Henry James have no museumised houses associated with them (Trubek, 2011) although James' former home, Lamb's House in the historic town of Rye in England, commemorates him and other authors who inhabited the Georgian house and wrote in its Garden Room.

The vast majority of writer's homes are, of course, associated with deceased authors who no longer need an earthly abode, with living writers generally only being encountered at literary festivals and book signings. However, some writers are involved in the recreations of house museums dedicated to their lives – Gabriel Garcia Márquez and his family assisted with the recreation of his childhood home in Columbia and Astrid Lindgren acquired hers in Vimmerby in Sweden in the 1960s, restoring it to its original décor and furnishings. This authentic home is now owned by Lindgren's nieces but can be visited by guided tours (Bom, 2015). One of our most famous and successful living authors, J.K. Rowling, creator of the *Harry Potter* series of books recently bought her childhood home, Church Cottage, a mid-Victorian revival house in the village of Tutshill in Gloucestershire where she lived from the age of 9 to 18. The house has several features which have been said to have inspired the author, such as a cupboard under the stairs, a trap door and on a wall in the author's old bedroom can be found a scribble stating, 'Joanne Rowling slept here circa 1982' (BBC News, 2020). The cottage has long attracted Harry Potter fans who can only stand outside and stare but perhaps this too will become a writer's house museum in the future.

Presenting the writer's house

Despite the great variety of authorial homes that have been preserved over the decades and across the globe, Robinson and Anderson suggest that there is something rather similar about them all – inside, the spaces where writers worked are 'often located within libraries reflecting their craft … Desks and chairs are surrounded by accumulations of collected material from travelling as sources of on-hand knowledge. Restful, bucolic views without and natural light prevails' (2002, p. 16). Although this set-up does feel very familiar to anyone who is a frequenter of authorial homes, there are of course many exceptions to this approach to presenting domestic settings. Linda Young explains that as with all museums created out of the homes of celebrated individuals, the writer's house may be presented in a number of different ways, utilising a variety of artefacts both authentic and reproduced. The house might be displayed to the public as it was found or with some stabilization. It may be restored to a known condition or recreated. The house could be presented as empty, partially, or fully furnished and these furnishings may be original or recreated using appropriate historic materials or reproductions. Finally, one or more historic periods may be presented within the one house (Young, 2019, p. 167). This complex array of different settings requires a variety of interpretive responses. Some authors' houses take a traditional museological approach and simply display their artefacts *in situ* in vitrines, either in the house itself or in attached visitor centres or exhibition spaces. This time-honoured approach to preserving precious items and creating a sense of aura around them by setting them apart for contemplation is still used but, more commonly, houses are set up as domestic spaces, using authentic artefacts, furnishings, and personal items laid out in a naturalistic manner to suggest current usage. Thus, a pen may be laid on a desk as if it has just been put down, or a hairbrush set on a nightstand, beside a half-read novel. The spartan interior of Hans Christian Andersen's childhood home in Odense is tiny with only a few objects on display such as the tools of his father's shoe-making trade and a half-burnt candle. This presentation may be minimal, but it very effectively denotes the cramped poverty which, it is said, allowed the young Hans' imagination to soar.

Decisions are taken about how much interpretive material to provide and how to supply it, as can be seen in Box 4.1 which explores the processes by which La Maison de Colette was set up between 2011–2016. Labels and information panels can disrupt the sense of an authentic, homely setting and such information can now be conveyed digitally via apps and Quick-Response (QR) codes, thus leaving the rooms free of any contemporary intrusion. Karen Blixen's Nairobi home has a wealth of personal objects laid out in a naturalistic manner with interpretation at a minimum, but the cordons placed around the desk, dining table, and beds create a barrier which certainly remind the visitor that they are in a museum.

Another key curatorial decision is whether to provide guided tours, self-guided tours, printed materials, or mechanisms to manage visitor flow, or to allow visitors to roam freely. Each approach has its benefits and limitations, and a knowledgeable and engaging tour guide can be a huge asset in presenting a writer's home but can also be a source of dissatisfaction, particularly as tour groups may be made up of many different types of visitors with competing needs. Some sites utilise costumed actors to present the tour, creating a character and conveying a particular perspective on the author themselves. One very successful example of this is the Housemaid's Tour at Charles Dickens' House in London. The maid opens the door to visitors and guides them around the house while the master is out. Keeping within character, the maid conveys gossip and factual details of the years that the Dickens family lived in the house and encourages interaction from tour group members. It is telling that these sorts of theatrical interpretations usually rely on a character contingent to the author rather than the writer themselves – Trubek suggests that this is because a true writer's abode is a house 'in which not much happens. Not one … has a reenactor scribbling furiously at his desk to entertain the crowds' (Trubek, 2011, p. 147). The act of writing is not a spectacle, and little can be made of it from an interpretative perspective although there are some examples of houses that present models of the author sitting at their desk or engaged in other writerly activities, for example in the Charles Kingsley Museum in Clovelly, North Devon, a model of the author sits at his desk writing a letter to his bride-to-be with a blanket over his shoulders while the English actor Joss Ackland recites his famous poem *The Three Fishers.* The author's voice can be used to good effect as is the case in Astrid Lindgren's home where recordings of the author reading extracts from her work and telling visitors about her childhood spent in the house enliven the guided tour (Bom, 2015).

However, the curatorial interventions explored above may not necessarily lead to the most affecting communion with an author at home. In her essay on *Great Men's Houses* (1932), Virginia Woolf compares the experience of visiting the Carlyles' well accoutred home in Chelsea with that of Keats in Hampstead. The couple's domestic set-up reminds Woolf of the labour and trials of their married life which seems to hang around the water pump and the cracked hip bath. In Keats' silent, unfurnished rooms in North London, however, 'all the traffic of life is silenced. The voice of the house is the voice of the leaves brushing in the wind; of branches stirring in the garden. Only one presence – that of Keats himself – dwells here' (Woolf, 1980, p. 22).

The practice of heritage interpretation was introduced by Freeman Tilden in 1957 in his publication *Interpreting Our Heritage.* His definition of the practice as 'an educational activity which aims to reveal meanings and relationships through the use of original objects, by first hand experience, and by illustrative media, rather than simply to communicate factual information' (Tilden, 2007, p. 33) has influenced the presentation of cultural and natural heritage sites until recently when divergent views on this approach have emerged. These challenges recognise the role of the visitor in the co-creation of the experience, the existence of multiple perspectives, including those of the host community, and finally the need for interpretation to embrace digital technology (Faith, 2022). These approaches are increasingly being applied to biographical sites and to other literary locations as we shall see in subsequent chapters.

Box 4.1 Case study: Establishing a new writer's house – La Maison de Colette

The case of the recently established La Maison de Colette illustrates the various decisions that must be taken when reconstructing and interpreting a new writer's house. The house in which the much-loved French author Colette (Sidonie-Gabrielle Colette) was born in 1873 and lived until the age of

18, recently joined the ranks of France's National Federation of Writers' Houses and Literary Heritage when it was first opened to the public in 2016. The house is in the village of St-Sauveur-en-Puisaye in Northwest Burgundy and was purchased in 2011 by the Society of the Friends of Colette who organised a high-profile fundraising gala to secure it. The Director, Frederic Maget, assembled a team of restoration specialists who decided to restore the house to the period of the author's childhood occupancy (1873–1891). The restoration took five years and cost one and a half million Euros (Engelking, 2018, 2019). Currently about 30% of the objects on display are authentic Colette family pieces and the lighting, décor, and soft furnishings were recreated to reflect the house as she would have known it. This reconstruction extends to the planting in the garden (which was the domain of Colette's mother Sidonie) and to the recreation of smells within the house. Nineteenth century editions of books which the family owned have been sourced to grace the bookcases.

The house is presented as if it is a living, domestic space – there are no signs, barriers, or other museological accessories to detract from this. The table is set for dinner and a game of dominoes is in full swing. Interpretation is provided via tour guides who use storytelling and excerpts from Colette's work to bring the rooms to life, with the entering of each room being described as the turning of a page in a 'book house' (La Maison de Colette, 2022). The groups are kept intentionally small so that the guide can tailor the interpretation to the characteristics and knowledge of the group members. The tour groups are led through the rooms and into the garden before the tour ends in the bookshop where it is hoped that a visitor will be inspired to buy one of her novels such as the *Claudine* stories (1900–1904), *Gigi* (1944), or *The Vagabond,* (1910). The intention of this interpretation is to revive interest in the author, whose works used to be required reading in French schools. As Colette didn't actually write any of her books in the house, the Director wanted to demonstrate the influence that her childhood spent there had on her later career as a writer. As the website for La Maison de Colette states, 'better than a simple writer's house, Colette's birthplace is the threshold of her literary creation' (La Maison de Colette, 2022).

La Maison de Colette has been successful in terms of visitor numbers, with 36,000 passing through the house between 2016 and 2019. Along with the house visit, the site also offers lectures on the work of Colette and other women writers, literary festivals, and cookery workshops where the favourite dishes of the author's childhood can be recreated.

The visitor experience of the house

Having examined the evolution of the writer's house concept, the means by which they come into being, and the different ways they can be presented, we can now turn our attention to the question of why visitors would wish to visit such a site and what they may get out of the experience. We have already noted that writers' homes are perhaps the most emotive and immediately engaging of literary tourism sites and visitors can perhaps relate most closely to the domestic setting being offered as it closely mirrors their own experiences. Trubek (2011, p. 5) notes that, unlike artists who have galleries and musicians who have concert halls, authors have no true 'gathering place' for fans to convene and so we make a shrine of their homes to provide a meeting spot for like-minded pilgrims. These homes, where a writer lived and worked, was inspired, and perhaps wrote some of their best loved works then becomes a vehicle for many sorts of emotional responses from the visitor. Firstly, we have the intimacy and authenticity of the author's home – the *frisson* of standing where they stood and interacting with the genuine fabric of a place that may have remained unchanged since the author's time. The power of this is evident in the souvenirs that early literary visitors used to take away from their trips – a practice still undertaken by contemporary visitors. The crab-apple tree at Bidford that legend tells us sheltered Shakespeare after a heavy night of drinking became nothing but a stump by 1824 after successions of Stratford visitors took pieces away as a physical connection with the Bard (Ousby, 1990). Leaves and pebbles were collected by Victorian visitors with Wordsworth's Lakeland garden being particularly hit by those seeking plant souvenirs. The present author is not immune to such leanings – she has attempted and failed to propagate a cutting from the ancient wisteria tree on William Morris' Red House and pressed leaves from the creeper that adorns his Kelmscott Manor. The practice of leaving one's name or an inscription at an authorial site was also common practice. Signatures on walls and scratched onto windows became a feature until the visitor book was introduced as a means of collecting visitor sentiment without any further damage to the house. This allows the visitor an opportunity to leave something of themselves behind as well as providing useful feedback to site managers. Online review sites now offer a more remote version of the visitor book.

Harald Hendrix (2008, p. 2) describes the writers' house as a 'medium of expression and of remembrance.' It is the outward articulation of the authors' life and work, and this offers a focus for recalling what we know of the author and their works within their own domestic sphere. Thus, what the visitor is really looking for in a writer's house is the reincarnation of the writer themselves and perhaps an engagement with the creative spark which brought favourite novels into being. Maybe there is a hope that the spirit of the author resides in the rooms they frequented and in the objects that they used, and this is what draws visitors to these domestic shrines. Certainly, the presentation of many of these homes would suggest that this might be the case, although the ever-sceptical Trubek (2011) suggests that to go in search of an author in their erstwhile home is a 'fool's errand' (p. 5) as we will never find what we seek there, only empty rooms. Less starkly, Booth (2016) suggests that there is 'something literal-minded about touring a literary region or an authorial house' (p. 4) as if we can find something that lives within our imagination and is wholly fictional within the everyday confines of domestic space.

Despite this apparent futility, the concept of authenticity seems to be important in attracting the literary pilgrim to writers' homes – both in terms of seeing the real setting in which the author lived and wrote and being able to engage with the concrete remains of the past which can take us back in time (Hendrix, 2008). Booth (2016) suggests that this time travel is not conceptualised as a history lesson *per se* but as an expression of personal heritage. Therefore, the writers' house does not instruct us in eighteenth century wall coverings or Victorian table settings, but instead allows us to imagine that we ourselves can go back and imaginatively inhabit the house and perhaps encounter its former incumbent. We can, in fact, imagine for a time that the house is our home, and in some circumstances, it may even be possible to physically inhabit an author's house as we can see in the examples of literary homes currently available as holiday rentals in Table 4.1. Spending a holiday in a writer's home and exploring their local haunts may provide the fullest literary tourism experience for the pilgrim, or an appealing opportunity for writers to gain inspiration from an authentic and rarefied setting.

The many uses to which a visitor might put an authorial home render these settings as multi-faceted sites of meaning. They are settings for our emotional engagement with a favourite writer and their work, perhaps providing 'backdrops for recollection and imagination' as Robinson and Andersen (2002, p. 17) suggest. As imaginative spaces, the visitor can create what they wish within the walls – some will be placing themselves at the writer's desk, channelling the spirit of the deceased author. Others will be examining the fabric of the building and its environs looking for clues to imaginative setting and action. And inevitably others will be enjoying a pleasant day out with the hope of a cream tea and a reviving visit to the gift shop. In her article on French writers' houses, Bonniot-Mirloup (2016) refers to these spaces as heterotopias – places with different layers of meaning for visitors and with diverse relationships to other spaces, both of which are not always apparent to the observer. Thus, the writer's house will have a historical dimension, displaying the architecture and furnishings of the writer's time, alongside a layer of spirituality wherein the visitor might feel the presence of the lost artist. Both dimensions are bound up in the domestic scene which is being presented and again that scene has various temporalities – the contemporary visit, the author's era being evoked, and perhaps even the time of the visitor's childhood as nostalgia moves them back and forwards in different times of their lives.

One feature of writers' houses which seems to be particularly potent in channelling the visitor's gaze is the opportunity to look through windows and into mirrors, or to imagine peering through magnifying glasses or spectacles belonging to the writer. Nicola Watson gives a fascinating analysis of the importance of glass within writers' homes in her detailed study on *The Author's Effects* (2020). She comments on the ubiquity of authors' spectacles in writers' house displays (we can, for example, view the spectacles of Virginia Woolf and the magnifying glass of Thomas Carlyle) and these intimate objects, through which a writer has viewed their world, appear to be a direct

Table 4.1 Authors' homes available to rent

Rental property	*Literary connection*
Barnhill Farm, Isle of Jura, Inner Hebrides, Scotland	This isolated farmhouse on the remote island of Jura was George Orwell's rented residence between 1946 and 1949. Here he wrote his novel, *1984*. The farmhouse is off grid with its own generator and visitors can book weekly stays in the house which has not changed since Orwell's day.
Fleming House, Goldeneye Resort, Jamaica	Fleming House on the Goldeneye Resort in Jamaica is where Ian Fleming wrote all the James Bond novels. He bought the 15-acre estate in 1946 and designed the villa himself. Guests can now rent the secluded three-bed beach front property.
Greenaway Apartment, Devon	Visitors can stay in a four-bedroom apartment that forms part of Greenway, Agatha Christie's former Georgian holiday home in Devon. The house is full of artefacts related to the Christie family and she set *Dead Man's Folly* in the house, boathouse, and locality which are accurately described in the novel.
43 Cloth Fair, London	The Georgian house at 43 Cloth Fair in London's Smithfield was home to poet John Betjeman for 20 years from 1954. He lived in two rooms above a shop and this one-bed apartment is available for holiday lets. It is decorated in the poet's personal style, which includes William Morris wallpaper.
Ted's House, Aspinall Street, Mytholmroyd near Hebden Bridge, West Yorkshire	The modest two-bedroom end of terrace cottage where poet laureate Ted Hughes spent his childhood can be rented. A desk has been installed in Hughes' bedroom to accommodate visiting writers who may feel inspired by the surrounding moors, as the poet was.
Cottage at Hilltop, Lake District, Cumbria	This two-bedroomed apartment forms part of the large Lakeland farmhouse that was Arthur Ransome's last home. The house dates from 1680, is situated on Lake Windermere, and a plaque commemorates its literary connection. *Swallows and Amazons* was set in this region.

route to the embodiment of the author. Often found on desks, they also form part of the writers' tool kit. Similarly, mirrors which have reflected the image of the writer many times in their lives seem to be potent objects. We can see our reflection in the same glass that reflected our beloved artist providing a reunion within the mirrored recesses. Windows that provide viewpoints which the author would have seen daily can also provide a moving experience for the visitor. This author found that she was less affected by seeing the tiny table where Jane Austen wrote her novels than looking through the window of the small bedroom that she shared with her sister Cassandra which gives onto the yard and outbuildings which have not been altered since her occupation. This unchanged view which she would have witnessed every day and in all weathers perhaps allows a greater insight into her life than any other aspect of the house at Chawton.

Studies and writing rooms

Writers' homes present a variety of rooms to the visitor, for example at Louisa May Alcott's home, Orchard House, the tour includes parlour, kitchen, dining room, her bedroom, her sister May's bedroom, her parent's bedroom, the nursery, and her father's study. Out of all these well-furnished rooms (Orchard house boasts that 80% of its artefacts are genuine Alcott belongings), Louisa's bedroom with the small half-moon shaped desk by the window, which her father made for her and on which she wrote her novels, has the most meaning for visitors. The novelist's desk and its associated writing equipment are part of the 'sacred hagiography' of the writer (Robinson and Andersen, 2002, p. 10) and will always provide a special lure for visitors. This piece of furniture may be tucked away in a bedroom, taking pride of place in a study or library or even, in the case of Mark Twain's sumptuous house in Hartford, Connecticut, in the billiards room. The surface over which the author's pen moved, creating the texts that made them famous, draws the visitor close and gives them clues about the physicality of the writing process. Longfellow's standing desk in his study in Cambridge, Massachusetts suggests an urgency and energy in the act of writing whilst the board resting across the arms of a well-padded easy chair in Roald Dahl's writing hut denotes a much more relaxed mode. Authors in search of more solitude than the house could afford might have their writing space hidden away in a shed or building away from the daily demands of domesticity. On the desk will sit pens, blotters, spectacles, and perhaps even the most sacred of objects – the unfinished manuscript with every pen stoke and blot demonstrating the humanity of the author. Other domestic items such as coffee pots and teacups may complete the picture. It is this *mis en scene* that most often prompts visitors to feel that the author has just put down their pen and stepped out of the room for a minute. Their presence seems to hang around the tools of their trade as a study of Tripadvisor reviews of writers' homes in the USA and UK reveals (MacLeod, 2021). In this study of 1200 detailed reviews, visitors repeatedly mentioned the significance and value of seeing artefacts associated with the writer's craft – pens, inkstands, and desks. They also reported sensing the spirit of the deceased author, whether that be in terms of the haunting of the rooms by the ghost of the author, a feeling of residual aura, or a feeling that the writer had just left the room and would be back at any moment.

Self-contained writing rooms or sheds where the author has hidden themselves away from the everyday activities of the home also attract envious attention. These 'detachable chapels or reliquaries of writing' (Booth, 2016, p. 57) underpin the Romantic solitude of the writers' art and in these settings, we can imagine the author devoting themselves entirely to this work. George Bernard Shaw's tiny writing shed at the end of a large garden allowed him to escape the constant ringing of the telephone and had the added advantage of being on a turntable so that it was always warmed by the sun's rays. Virginia Woolf's writing lodge to which she commuted with the regularity of an office worker from her home, Monks House in Sussex, was situated in the garden with open views to the countryside. Here she found peace, sometimes sleeping in the lodge in the summer months with her husband Leonard's apple stores hidden away in the attic above. Roald Dahl's writing hut was meticulously organised so that everything was within reach of his chair. The interior of the shed was moved from the family home to the nearby Roald Dahl Museum and Story Centre in Great Missenden and is a popular exhibit, still holding a lingering smell of Dahl's tobacco. Perhaps the most famous writer's shed of them all was of course Henry David Thoreau's cabin at Walden Pond near Concorde, Massachusetts in which he lived for two years, writing about his experiences in nature in his book *Walden: or Life in the Woods* published in 1854. The fame of this book and its author's early death led to many pilgrims making their way to Walden Pond where the site of the cabin is now marked by granite posts. A replica cabin was built in 1985 showing the simplicity of the dwelling with bed, wood burning stove, and desk.

Work-orientated visitors who may not want to rent an entire author's home to holiday in may instead rent a writer's study in which to find inspiration. The birthplace of Ernest Hemingway in Chicago is offering Society members the chance to 'work from Hemingway's for a day. Write email masterpieces at his desk. Zoom your co-workers from his couch. Do inspiring research in his library. And when you're ready for a work-day happy hour, enjoy Hemingway's favorite cocktail at noon' (Hemingway Birthplace, 2022). Prior to the Covid-19 pandemic of 2020 onwards, aspiring writers could also rent Mark Twain's library in his house museum in Hartford, Connecticut as a temporary workspace.

Artefacts

The writer's house is arguably the 'apogee of literary tourist sites' (Watson, 2006, p. 90) as it holds the strongest biographical link to the author. And within the walls of the authorial home are objects once owned by the writer, providing for some visitors a direct link across the years. We have already considered the particular allure of the desk, the study, and the tools of the writer's craft but other belongings also hold totemic power. These items include domestic articles which demonstrate the author as centre of the household as well as gifted artist. Other objects such as clothing and jewellery and other more intimate objects can sometimes create a sense of voyeuristic discomfort in the viewer. And finally, we also find physical relics of the author – locks of hair, death masks, or models of body parts. Aislinn Hunter in her thesis on *The Reading of Resonant Things and Material Encounters in Victorian Writers' Houses/Museums* (2015) refers to these powerful objects as 'autographic' in that there is a 'belief that part or all of the object's "life" parallels the life of the person with whom the object is associated; a shared biography usually supported by some form of provenance' (p. 52). A close connection to the author aligned to verified authenticity gives these objects their distinct resonance. Although they would still be powerful within a decontextualised vitrine, these artefacts gain additional glamour within the room settings of the writer's home where the domestic surroundings 'appear to bless the personal effects, or indeed any effects, taking the tourist experience beyond that of a museum' (Robinson and Anderson, 2002, p. 18).

Some objects within writers' homes have become well-known in their oddity and intimate nature and have entered popular culture. Flaubert's parrot, Petrarch's mummified cat, Emily Brontë's couch, and Emily Dickinson's white dress are all examples of such renowned artefacts. Items which were owned and used by the author, and which also appear in their works, are particularly prized – Louisa May Alcott's red mood cushion which she used to show her family what temperament she was in can be viewed in the parlour at Orchard House and is a highly personal object in its own right. However, given that Louisa's much-loved heroine Jo March also had such an accessory which appears in *Little Women* (1869), it attains even more power. Jo's friend Laurie was very much attuned to the vagaries of this cushion and 'if "the sausage" as they called it, stood on end, it was a sign that he might approach and repose, but if it lay flat across the sofa, woe to man, woman, or child who dared disturb it' (Alcott, 2020, p. 252). The rather ordinary, dusty red bolster therefore has both an intimate connection to the life of the author and an imaginative connection to her most beloved creation eliding the biographical with the imaginary with potent effect. The tiled fireplace in Virginia Woolf's bedroom at Monk's House which shows an image of a boat approaching a lighthouse was painted by Woolf's sister Vanessa Bell in recognition of her success with her 1927 novel *To the Lighthouse* and is signed 'V.W. from V.B. 1930.' This very personal domestic detail reflecting Woolf's imaginative landscape makes this room a very compelling site for her fans. Writers' homes and possessions which have this direct link to their creative works are therefore perhaps the most affecting for visitors as we will see in the following chapter.

We have already seen that the tools of the writer, along with associated domestic items, are prized by visitors for their immediate links to both the creative genius of the artist and their everyday experience and are thus used by museum curators to create a domestic scene and a sense of the house as lived space. Writers' houses also display more intimate items associated with the author such as personal accessories and clothing. Watson (2020) analyses a range of displayed clothing suggesting that curators use these items to delineate the lost body of the author, with such exhibits often giving rise to visitor comments about the height, shoe size, or waist measurements of the previous owners. The clothing effectively brings the absent body back into the present, often creating a rather ghostly stand-in for the lost author as is the case with Emily Dickinson's white house dress displayed on a mannequin in her bedroom in the Dickinson House Museum in Amherst, Massachusetts. Clothing can of course tell us much about how a writer wanted to project themselves to their public, and visitors to George Bernard Shaw's house at Shaw's Corner in Ayot St Lawrence, Hertfordshire can view one of his famous Jaeger woollen capes. He bought the first in the 1890s and was often photographed wearing them. He was specifically drawn to these garments as he had adopted Dr Jaeger's 'Sanitary Woollen System' which underlined the potential for natural unbleached wool to aid the escape of perspiration if worn next to the skin (National Trust, 2022b). Thus, this garment both demonstrates Shaw's chosen public persona and speaks of the functions of a body no longer with us.

Authentic objects associated with the author's career as a writer are also prized and take centre stage. First editions, literary awards, favourable reviews, and other symbols of esteem are presented as part of the authorial biography. Back at Shaw's Corner, visitors have the unique thrill of being able to see the playwright's Nobel Peace Prize for Literature (1926) alongside his Oscar (he won Best Screenplay for *Pygmalion* in 1938). The glamour of these glowing, golden awards sits slightly uneasily in the small study of a quiet Hertfordshire retreat.

Bodily items

The most personal items to be found within writers' houses and other collections are of course relics taken from of the author's person. Like the grave, which we consider later, such relics allow the visitor to gain the most immediate connection with the physicality of the authorial body. These artefacts may be pieces of bone or hair taken from the body at death or during exhumation, or they may be casts taken of different body parts such as death masks. When viewed along with busts and portraits which are commonly found in house museums and which help to contextualise the relics, these artefacts allow the visitor to gaze on the corporeal remains, gaining an intimate insight into their physical characteristics.

Locks of hair were exchanged as mementoes during life and were commonly distributed to mourners after death as funerary items, often incorporated into a piece of mourning jewellery. These small but intimate fragments can be found in several collections and writers' homes such as the Keats-Shelley House in Rome. The Leigh Hunt hair collection (gathered by the Romantic poet himself) resides in the Ransom Centre at the University of Texas and includes samples which continue to reflect the physical attributes of the writers – a lock of Mary Shelley's hair is still auburn, Elizabeth Barrett Browning's retains its curl, and the older Wordsworth's sample is thin and greying. Incredibly, these relics of the authorial body could be handled by visitors up until the 1990s. In 2021, a pair of lockets holding the hair of Charles Dickens and his sister-in-law, Mary Hogarth was acquired and displayed by the Charles Dickens Museum in London. Charles was very fond of Mary who died in his house at the age of 17. The intimacy of these exchanged locks of hair could be viewed by visitors with some prurience although the Museum preferred to use the artefacts to explain the influence of Mary's death, which haunted Dickens' work in the illness and death of young female characters such as Rose Maylie in *Oliver Twist* (1838) and Little Nell in *The Old Curiosity Shop* (1841).

Authors' skeletal remains do not always remain safely buried after their demise and there are several accounts of exhumations, whether due to contested 'ownership' (in the case of Dante's remains in Ravenna) or theft, as happened to John Milton, some of whose remains were sold off by the verger while they were being temporarily moved from their resting place in St Giles' church in the City of London in 1790. Some of Milton's bones, teeth, and hair were bought back and re-interred in his grave. A lock of hair in the collection at Milton's Cottage in Buckinghamshire may have come from this original plundering (Watson, 2006). Less dramatic was the opening of Robert Burns' grave to allow for the burial of his wife Jane Armor beside him. During this process, Burns' skull was temporarily removed, and a cast made of it which has subsequently resided in the Burns Museum. Recently, new forensic technology, based on a scan of the skull has allowed a reconstruction of the poet's head to be created. This too is on display at the museum and gives visitors a lifelike three-dimensional impression of the poet's countenance, augmenting the famous Romantic portrait of the author by Alexander Nasmyth. This image, in the permanent collections of the Scottish National Portrait Gallery, has become the *de facto* official portrait of the Bard, adorning countless tea towels and shortbread tins over the years.

Casts taken of body parts also allow for a physical engagement with the corporeal writer whether they be death masks modelled on the deceased's face for the sake of commemoration or to assist sculptors in modelling busts. Death masks of Keats are on display in his homes in Rome and London and one recently sold to a private owner for £12,500 (Lloyd, 2021). A cast taken of Thomas Carlyle's hands while he was still alive is on display in his home in Cheyne Row. Sir Joseph Edgar Boehm took the cast in 1874 while making a statue of the author. The white disembodied fingers are lightly clasped, and a few veins stand upon on the backs of the hands, giving a ghostly but intimate insight into the hand that held the pen.

Haunts

The term 'haunts' comes, of course, from William Howitt's very influential publication, *Homes and Haunts of the most Eminent British Poets* (1875) and refers to the places that authors frequented as an extension to the home. Bonniot-Mirloup (2016) suggests that an augmentation of the literary theme occurs when the literary site is extended into the wider environs surrounding the writer's home. These environs may include visits to the homes of friends, local businesses, places of work, and favourite walks. In the London borough of Bexleyheath, home to William Morris' Red House, there is a marker which looks remarkably like a gravestone informing passers-by that this was the route that Morris took every day on his journey into London via Abbey Wood station. For the William Morris society at least, a literary commute is worthy of commemoration.

The literary landscape surrounding Haworth, forever synonymous with the Brontë family, is replete with haunts. At its core is the Parsonage and the Reverend Patrick Brontë's Haworth Parish Church. This church was the scene of Charlotte's marriage and is the site of the graves of all the Brontë family but Anne. The nearby Haworth Old School Room was built by Patrick, and his daughters all served as teachers there. Above the village, visitors can access the moors, where markers of the family can be found everywhere from the Brontë Falls, the walking route named the Brontë Way, and a plaque affixed in 1965 by the Brontë Society on the lonely ruins of the farm at Top Withens telling walkers that this was, in fact, not the inspiration for Emily Brontë's *Wuthering Heights* (1847) although disappointed pilgrims are assured that 'the situation may have been in her mind when she wrote of the moorland setting of the Heights.' MacLeod (2021, p. 1214) suggests that 'this physical and imaginative extension of the literary milieu reflects the concept of authorial homes and haunts and demonstrates the pervasiveness of literary connections.' The concept of the literary trail makes full use of the wider associations that make up the literary landscapes as we shall see in Chapter Eight.

Robinson and Andersen note that 'somewhat removed from the idea of "home" for the writer are those places of transitory association: places where the literary figure dined, was entertained, slept, or made some remembered stop in his or her passage through life' (2002, p. 18). An interesting sub-group of haunts is therefore the literary hotel where writers sought refuge, peace, and inspiration to write, away from the demands of home. Even more compelling for the literary tourist are those hotels which are still open to the paying public and where visitors might sleep in the same room as a favourite artist. Raffles Hotel in Singapore is possibly one of the most famous hotels to boast literary connections, with writers such as Conrad and Kipling enjoying the colonial art deco palace which opened in 1887. The hotel has recently launched a writers' residency programme and opened a Writers' Bar in honour of its heritage. Agatha Christie loved to write whilst staying in hotels and is said to have created *Murder on the Orient Express* (1934) in Room 411 of the sumptuous Pera Palace Hotel in Istanbul. The Agatha Christie Room still has its original antique furniture, a library of Christie's books published in several languages, and a replica of the typewriter she used. Hotels featured heavily within her plots – in her short story *Problem at Pollença Bay* (1936), the action takes place in the fictional Pino d'Oro Hotel in Pollença on the northeast coast of the Balearic Island of Mallorca. In the tale, the hotel is described as 'standing on the edge of the sea looking out over a view that in the misty haze of a fine morning had the exquisite vagueness of a Japanese print' (Christie, 1994, p. 4), a characterisation which aptly fits the Illa d'Or Hotel where she stayed in 1932. The hotel is now a feature on an Agatha Christie literary trail on the island (see Fig. 8.1). The Mandarin Oriental in Bangkok hosted John le Carré and Somerset Maugham amongst other literary guest and now boasts an Authors' Wing. In Paris, the modest pension where Oscar Wilde lived out his last two years is now one of the city's smallest five-star hotels. L'Hotel has a large collection of Wilde related memorabilia including a letter regarding an unpaid hotel bill.

Sometimes the biographical link to an author may be rather tenuous – the Hobbit Boutique Hotel in Bloemfontein, South Africa is a J.R.R. Tolkien-themed hotel celebrating the fact that the author was born in the town in 1892 and lived there until he was three. The hotel incorporates parts of the original Tolkien birthplace (the original house was lost to floods in the 1920s) and a plaque announcing the site of his birth. Hobbit fans can enjoy the themed suites and bar and venture forth to discover the bank which Tolkien's father managed and the cemetery in which he was buried. Literary hotels are further examined in Chapter Seven where places of literary reputation are considered.

The restaurants, cafés, and pubs where authors socialised or perhaps worked can also become part of the literary tourism landscape and are incorporated into tours and trails as we shall see in Chapter Eight. Some of the more convivial authors and poets bestowed their blessings on a wide range of such hostelries and others, such as J.K. Rowling, famously used cafés as ersatz studies. The cafés where she began writing her Harry Potter series in the 1990s are popular fan pilgrim sites with visitors perhaps hoping to be similarly inspired. Commemorative plaques mark the sites, photographs of the author working on site are displayed and, in one café, fan graffiti adorns the bathroom.

The grave

We have seen that a deceased author's house, no longer being required to shelter the artist, can be turned into a visitor attraction and, along with other significant sites and haunts, can contribute to the tourism offering in both urban and rural settings. Robinson and Andersen suggest that dead authors offer advantages to the tourism industry as they 'provide us with graves and memorials to visit. These are tangible signatures of a writer's presence and commonly accessible to the public' (Robinson and Andersen, 2002, p. 19). In this section we will situate the habit of visiting writers' graves within the wider practice of cemetery visiting and consider the specific motivations and experience of the literary visitors at the graveside.

Travel to visit graves and cemeteries has a long history (Watson, 2006) and can be described as a form of dark tourism which is defined by Lennon and Foley (1996) as 'the presentation and consumption (by visitors) of real and commoditized death and disaster sites' (p. 198). Aside from graves, dark tourism sites might include battlefields, sites of execution, extermination, or accidental death, and the locations of disasters, whether natural or human made. The older term 'thanatourism' where the visitor is drawn towards 'actual or symbolic encounters with death' (Seaton, 1996, p. 234) is also associated with the practice of visiting graveyards and individual tombs, amongst other sites of suffering or death. These dark tourism sites can be conceptualised as a continuum from the lightest of dark sites (such as visitor attractions using decontextualised dark themes for entertainment), to the darkest settings, where authenticity is high, and the focus is on education and commemoration. Therefore, grave visiting is placed at the darker end of the spectrum (Stone, 2006).

Linked to the notion of dark tourism is the Romantic impulse to visit places representing the sort of gentle, picturesque decay found in ruins such as those of Tintern Abbey made famous by William Wordsworth's poem of 1798. These lines explore the poet's central themes of memory and the consolations of nature to which these ruins have returned. Historical graveyards provide a similar environment with ancient masonry, immortal yew trees, and ivy-covered tombstones leaning back towards the earth. These peaceful places also provide a sanctuary for wildlife. The stones themselves are a rich repository of social history, showing local occupations, family size, life spans, and reasons for death. Highgate West cemetery is an evocative, steeply wooded funerary landscape that is the model of a haunting, Gothic burial ground, north of the city of London (see Box 4.2). This cemetery (and the newer site to the east) houses numerous famous deceased which provide a point of interest, along with the various architectural styles and other historical detail. But the graveyard also encourages a more diffuse experience, with visitors finding that the environment of Romantic decay gives them cause to ponder on the fleeting nature of time, the fragility of life, and our own mortality. In her study of Budapest's National Graveyard, Kerepesi Cemetery, Pécsek (2015) considers cemetery visiting as part of a region's cultural tourism offering, representing local and national identity for city-break tourists. She notes that the Association of Significant Cemeteries in Europe have put together a themed route which is now a certified European Cultural Route which emphasises the value of these sites to European cultural tourism. Their enduring interest is also exemplified in websites such as Find a Grave which allows potential visitors to search over 210 million graves, cemeteries, and memorials, many of celebrated individuals, as we saw in Chapter One (Find a Grave, 2022). People who have an interest in war graves and battle memorials, whether from a specific family link or for historical purposes, can find many guided tours provided by specialist operators.

Box 4.2 Case study: Literary cemeteries

Cemetery visiting has long been a popular tourist activity for visitors keen to search out the specific graves of well-known individuals or to absorb the history that is represented in both the architecture and the inscriptions of ancient tombs and monuments. Cemeteries also provide a peaceful sanctuary where wildlife can thrive, and visitors can walk. Historical cemeteries are often included in the cultural tourism offering promoted by destination management organisations and maps, tours, and other forms of visitor interpretation are offered to augment the experience. The following three cemeteries are particularly popular as they are well known as the final resting places of many luminaries, including several much-loved literary greats.

Highgate Cemetery, London

Highgate Cemetery was established in the nineteenth century along with six other London cemeteries to deal with the lack of burial space in the capital. The first burial took place in Highgate Cemetery in 1839, with the East Cemetery extension opening in 1856. The cemetery, on a steep wooded hill, has a number of different architectural styles, including Tudor Gothic and Egyptian, and is most famous for being the resting place of Karl Marx. Literary figures buried here include George Eliot, Christina Rossetti, and Douglas Adams. An entrance fee is charged for the historic West Cemetery and self-guided and guided tours are available along with talks and workshops (for example workshops in stonemasonry). Burials continue in the cemetery at the rate of one per week.

Père Lachaise Cemetery, Paris

Established in 1804, Père Lachaise is the largest cemetery in the city of Paris, stretching over 110 acres of hilly terrain. Over 5000 trees provide a peaceful atmosphere, despite it being a major visitor attraction and indeed the most visited non-secular cemetery in the world. This fame is due to the number of famous individuals interred here as well as the cemetery's situation within a major tourist city. Guided tours are not provided by the cemetery, but local tour guides offer a variety of themed walks on the site. The cemetery provides a free map and an online interactive tour to guide visitors. Literary graves that can be visited include those of Colette, Gertrude Stein, Oscar Wilde, Marcel Proust, and Honoré de Balzac.

Sleepy Hollow Cemetery, Concord, Massachusetts

Sleepy Hollow Cemetery in Concord, Massachusetts is a 119-acre site first established in 1855 and designed as a rural cemetery inspired by Romantic ideas on the power of nature to console. The cemetery is laid out in an undulating landscape with naturalistic planting. Sleepy Hollow is on the National Registry of Historic Places and is a very popular visitor attraction for tourists who wish to see the graves of local residents Louisa May Alcott and her father Bronson Alcott, Ralph Waldo Emerson,

Henry David Thoreau, and Nathaniel Hawthorne. These literary graves are all situated in an area that has become known as 'Authors' Ridge.' Signage points visitors to this hilly area and a map marked with 'graves of interest' plus an interactive online cemetery tour are also provided to assist visitors.

Although cemetery visitation may therefore have dark aspects or be motivated by a historical or cultural interest, it is debatable whether visiting the graves of literary greats can be conceptualised in this way as the literary visitor is not in search of death or of historical detail *per se* but of the temporary revival of a very specific lost author. In his work *Travelling to Meet the Dead* (2012), Westover places literary tourism within a necromantic framework, where the tourist travels to commune with the dead or to attempt to raise their spirit. This impulse is perhaps strongest at the literary grave where the visitor is in the presence of the authentic remains, but it is also a means of experiencing a writer's home as we have already seen above, where the spirit or aura of the deceased is felt within their domestic setting. There is a long tradition of pilgrimage to visit the mortal remains of illustrious individuals and this is probably the oldest form of literary tourism. We have already seen in earlier chapters that Virgil's tomb near Naples attracted visits from 19 BC, Petrarch's tomb near Padua was a focus for literary tourists from the 1440s, and Shakespeare's grave and monument at Holy Trinity Church in Stratford were points of pilgrimage from the 1620s. The significance of writers' tombs to national culture is of course clearly exemplified in the establishment of the 'poetical quarter' (subsequently Poets' Corner) in Westminster Abbey which began with the burial of Geoffrey Chaucer in 1400. From then on, this site has been the focus of many a devotional journey to visit the resting places of Spenser, Tennyson, Dickens, Kipling, Hardy, Masefield, and many more (male) literary luminaries, women writers being far better represented in memorials than actual burials in the Abbey. France similarly honours its dead poets in the Pantheon in Paris where the tombs of Alexander Dumas, Emile Zola, Victor Hugo, and the ashes of Voltaire can be visited.

Visiting the graves of illustrious individuals is therefore an accepted means of paying respect to the departed as it offers a way of communing with authentic mortal remains. Although the graves of famous individuals from all walks of life will engender interest, the final resting places of authors and poets have long held a particular appeal as Washington Irving suggested in his *Sketch Book* (1820). Here he contrasts the generic curiosity or admiration that one might feel at the grave of the great and the good with a much 'kinder and fonder feeling' which may be experienced at the tomb of a poet or novelist. Visitors 'linger about these as about the tombs of friends and companions; for indeed there is something of companionship between the author and reader' (Irving, 1820, quoted in Ousby, 1990, p. 29). Thus, we might admire the achievements of a great general or political leader, but as readers, we feel a much closer relationship with an author as we have spent more of our time with them, reading their works, and seeing the world through their eyes. This personal relationship is also augmented by a physical relationship which proximity to the bodily remains of the artist provides.

William Godwin understands the power of the literary grave in his famous *Essay on Sepulchres* (1809) where he feels sorry for those who spend their time at the tombs of illustrious individuals such as George the Third, Alexander of Russia, or Kant when they could be consorting with Chaucer and Milton as 'they are not dead' (Godwin, 1809, p. 82). Godwin's essay is a proposal to erect simple monuments to all the celebrated dead, but in particular, the literary deceased. He also recommends the creation of an 'atlas of those who have lived for the use of men hereafter to be born' (Godwin, 1809, p. 112) – in effect a map and a catalogue of notable graves to inform contemporary and future travellers and to provide a useful record in case of war. He hoped that in encouraging the population to travel to visit notable graves (particularly literary ones), they would develop a stronger spiritual and imaginative life and consequently he suggested that this atlas and catalogue be created through public subscription as a national memorial. Godwin would have approved of the searchable database described above which now allows the sort of commemoration and access which his proposed atlas was to provide.

The closeness of the relationship between the reader and the deceased author suggests that the most convincing explanation for the practice of literary grave visiting is the pilgrimage concept which is a common trope in tourist studies (Matthews, 2009) as explained in the previous chapter. The literary pilgrim travels in search of the spiritual centre and shows a deep interest in all aspects of the author's life and work, but perhaps is most moved to visit their homes and their final resting places (MacLeod, Shelley and Morrison, 2018). It is common for commentators to suggest that, historically, the Enlightenment encouraged the replacement of saints and their trappings with writers and their homes and haunts (Ousby, 1990). More contemporary tourism pilgrimages are explained by the new post-secular society and, for those new-age spiritual travellers, the pilgrimage is 'important to the core identity of the tourist: it is highly valued and deeply meaningful' (Brown, 2016, p. 168). In her analysis of literary pilgrims at the grave of Jean Paul Sartre and Simone de Beauvoir in Montparnasse cemetery in Paris, Lorraine Brown noted that they were motivated by a desire for communion, the need to pay homage, to find space for reflection, and the impulse to leave offerings (Brown, 2016). The offerings included the commonplace flowers and plants, but metro tickets, notes, and cigarettes were also left behind, demonstrating the pilgrims' knowledge of the deceased as both authors smoked heavily. The notes were either personal messages of thanks or quotations from the works. Other examples of personalised offering at writers' graves include the shell offerings on John Betjeman's simple grave at St Enodoc churchyard on the Cornish coast in the southwest of England or the pens and pencils that are increasingly left at writer's graves. Douglas Adams, author of the *Hitchhiker's Guide to the Galaxy* (1979), has an impressive collection of writing implements left by fans at his grave in Highgate Cemetery in London.

Conclusion

Writers' birthplaces, homes, belongings, and final resting places have long been popular sites of pilgrimage for literary-inspired visitors, and they continue to be developed and to attract the contemporary tourist. In this chapter we have reviewed the places, sites, and artefacts associated with the lives and work of artists and available to literary tourist visitation. These biographical literary resources demonstrate the importance of authenticity, although the prevalence of reconstructed sites suggests a willingness to suspend disbelief if other markers of originality are present. To make a pilgrimage to such sites is to seek the departed author in their home or at their grave and to perhaps attain some of the inspiration that was visited upon them. Preserving and visiting the homes and graves of the illustrious is not an unusual practice but in the case of writers, there appears to be a closer communion between them and their reader-visitors as the practice of reading allows one to spend time with the author. Visiting their home or their tomb allows one to extend that time even further and helps us to place the well-known genius within their biographical setting.

In the following chapter we will move on from places which attract visitors because of their links with the life of the author to the sites and settings which are the imaginary creations of that writer. In this chapter, therefore, we will seek out real and imagined literary locations and characters which live in the imagination of readers and can increasingly be visited as part of the literary fiction tourism offer.

References

Alcott, L.M. (2020) *Little women*, Volume One and Two. First published 1868–1869. Chesterfield: Chesterfield Classics.

Barnard, R. (2002) 'Tourism comes to Haworth', in Robinson, M. and Andersen, H.C. (eds.) *Literature and tourism: Reading and writing tourism texts*. London: Continuum, pp. 143–154.

BBC News (2020) 'Harry Potter: JK Rowling secretly buys childhood home', *BBC News*, 14 April. Available at: https://www.bbc.co.uk/news/uk-england-gloucestershire-52286400 (Accessed: 12 January 2022).

Bom, A.K. (2015) 'Where fiction really exists: The staging and experience of authenticity in the literary icon-city', *Aktualitylet - Litteratur, kultur og medier*, 9 (2), pp. 85–97. Available at: https://tjddrsk.dk/aktualityt/article/view/111371 (Accessed: 24 June 2023).

Bonniot-Mirloup, A. (2016) 'Tourism and writers' houses, in between places and literature', *Via. Tourism Review*, 9, pp. 1–18. Available at: https://doi.org/10.4000/viatourism.808 (Accessed: 24 June 2023).

Booth, A. (2016) *Homes and haunts: Touring writers' shrines and countries.* Oxford: Oxford University Press.

Brown, L. (2016) 'Tourism and pilgrimage: Paying homage to literary heroes', *International Journal of Tourism Research*, 18 (2), pp. 167–175. Available at: https://doi.org/10.1002/jtr.2043 (Accessed: 12 May 2021).

Çevik, S. (2018) 'Interpretation of author-related literary tourism sites: Evidence from three sites in Turkey', *Journal of Tourism Research*, 20, pp. 118–136.

Christie, A. (1994) *Problem at Pollença Bay and other stories.* First published 1936. London: Harper Collins.

Compton, R. (2016) 'William and the lost tourist' in *William the Conqueror* (first published 1926), London: Macmillan Children's Books, pp. 93–118.

Dickens, C. (1999) *Nicholas Nickelby.* First published 1839. London: Penguin Classics.

Donaldson, C. (2013) 'Down the Duddon: Wordsworth and his literary pilgrims', *Literary Imagination*, 15 (2), pp. 186–209. Available at: https://doi.org/10.1093/litimag/imt042 (Accessed: 21 May 2021).

Engelking, T.L. (2018) 'Literary tourism and La Maison de Colette: An interview with Frédéric Maget', *The French Review*, 92 (1), pp. 184–197. Available at: https://doi.org/10.1353/tfr.2018.0094 (Accessed: 21 May 2021).

Engelking, T.L. (2019) 'Speaking for Colette: Literary tourism, storytelling, and La Maison de Colette', *Women in French Studies Special Conference Issue*, pp. 104–116. Available at: http://purl.flvc.org/fsu/fd/FSU_libsubv1_scholarship_submission_1578588962_299d28f (Accessed: 12 June 2022).

Faith, G. (2022) '*Evaluating World Heritage interpretation in online spaces and its potential to prime the development of eco-cultural tourism experiences (virtual and onsite): A case study on Pimachiowin Aki, Canada's first Mixed World Heritage Site*', PhD thesis, Ulster University, Coleraine, NI.

Find a Grave (2022) World's largest gravesite collection. Available at: https://www.findagrave.com/ (Accessed: 15 June 2022).

Godwin, W. (1809) *Essay on sepulchres; Or, a proposal for erecting some memorial of the illustrious dead in all ages on the spot where their remains have been interred.* London: Miller.

Hemingway Birthplace Museum (2022) Work from Hemingway's. Available at: https://www.hemingwaybirthplace.com/work-from-hemingways (Accessed: 18 June 2022).

Hendrix, H. (2008) 'Writers' houses as media of expression and remembrance: From self-fashioning to cultural memory' in Hendrix, H. (ed.) *Writers' houses and the making of memory.* Abingdon: Routledge, pp. 9–20.

Hoppen, A., Brown, L. and Fyall, A. (2014) 'Literary tourism: Opportunities and challenges for the marketing and branding of destinations?', *Journal of Destination Marketing & Management*, 3 (1), pp. 37–47. Available at: https://doi.org/10.1016/j.jdmm.2013.12.009 (Accessed: 2 February 2021).

House of the Seven Gables (2022) Nathaniel Hawthorne's birthplace. Available at: https://7gables.org/history/nathaniel-hawthorne-birthplace/ (Accessed: 10 July 2022).

Howitt, W. (1875) *Homes and haunts of the most eminent British poets.* 3rd edn. London: Routledge.

Hunter, A.P. (2015) '*Evocative objects: a reading of resonant things and material encounters in Victorian writers' houses/museums*', PhD thesis, University of Edinburgh, Edinburgh. Available at: https://era.ed.ac.uk/handle/1842/21021 (Accessed: 1 July 2023).

James, H. (2013) 'The birthplace' in *The Aspern Papers and other stories.* Oxford: Oxford University Press, pp. 157–208.

La Maison de Colette (2022) Colette's house. Available at: https://www.maisondecolette.fr/ (Accessed: 6 July 2022).

Lennon, J. and Foley, M. (1996) *Dark tourism: The attraction of death and disaster.* London: Continuum.

Lloyd, A. (2021) 'Writ in water, preserved in plaster: How Keats' death mask became a collector's item', *The Guardian*, 26 February. Available at: https://www.theguardian.com/books/2021/feb/26/how-keats-death-mask-became-a-collectors-item?msclkid=5a244002c79e11eca594d99c8620280d (Accessed: 3 November 2022).

Lowe, H.I. (2012) *Mark Twain's homes and literary tourism.* Columbia: University of Missouri Press.

Mackay, P. and Pittock, M. (2011) 'Beyond text: Burns, Byron and their material cultural afterlife', *The Byron Journal*, 39 (2), pp. 149–162. Available at: https://muse.jhu.edu/article/463116 (Accessed: 5 November 2022).

MacLeod, N. (2021) '"A faint whiff of cigar": The literary tourist's experience of visiting writers' homes', *Current Issues in Tourism*, 24 (9), pp. 1211–1226. Available at: https://doi.org/10.1080/13683500.2020.1765996 (Accessed: 18 February 2022).

MacLeod, N., Shelley, J. and Morrison, A.M. (2018) 'The touring reader: Understanding the bibliophile's experience of literary tourism', *Tourism Management*, 67, pp. 388–398. Available at: https://doi.org/10.1016/j.tourman.2018.02.006 (Accessed: 20 February 2022).

Matthews, S. (2009) 'Making their mark: Writing the poet's grave', in Watson, N.J. (ed.) *Literary tourism and nineteenth-century culture.* Basingstoke: Palgrave Macmillan, pp. 25–36.

MIRE (2022) Association of Hungarian Literary Memorial Places. Available at: http://www.mire.hu/en/about (Accessed: 4 September 2022).

Nakamura, Y. (1990) 'The significance of fiction in Henry James's "The Birthplace"', *Studies in English Literature*, 66 (2), pp. 255–269. Available at: https://doi.org/10.20759/elsjp.66.2_255 (Accessed: 1 July 2022).

National Trust (2022a) Hardy's cottage. Available at: https://www.nationaltrust.org.uk/hardys-cottage/lists/explore-the-cottage (Accessed: 29 January 2022).

National Trust (2022b) Bernard Shaw's Jaeger cape. Available at: https://www.nationaltrustcollections.org.uk/object/1275501 (Accessed: 23 March 2022).

Ousby, I. (1990) *The Englishman's England: Taste, travel, and the rise of tourism.* Cambridge: Cambridge University Press.

Pearce, P.L., Morrison, A.M. and Moscardo, G.M. (2003) 'Individuals as tourist icons: A developmental and marketing analysis', *Journal of Hospitality & Leisure Marketing*, 10 (1–2), pp. 63–85. Available at: https://doi.org/10.1300/J150v10n01_05 (Accessed: 4 February 2022).

Pécsek, B. (2015) 'City cemeteries as cultural attractions: Towards an understanding of foreign visitors' attitude at the national graveyard in Budapest', *The Central European Journal of Regional Development and Tourism*, 7 (1), pp. 44–61. Available at: https://www.deturope.eu/contents/det/2015/01.pdf#page=45 (Accessed: 15 February 2022).

Pittock, M. and Ambroisine, J. (2019) *Robert Burns and the Scottish Economy.* Glasgow: University of Glasgow.

Robinson, M. and Andersen, H.C. (2002b) 'Reading between the lines: Literature and the creation of touristic spaces' in Robinson, M. and Andersen, H.C. (eds.) *Literature and tourism: Reading and writing tourism texts.* London: Continuum, pp. 1–38.

Rosenthal, M. (2008) 'Shakespeare's birthplace in Stratford: Bardolatry reconsidered', in Hendrix, H. (ed.) *Writers' houses and the making of memory.* Abingdon: Routledge, pp. 39–52.

Sabre, C. (2017) 'Between literary tourism and contents tourism: Marcel Proust and his Madeleine as a tourist attraction', *International Journal of Contents Tourism*, 2, pp. 1–11. Available at: https://contents-tourism.press/wp-content/uploads/2017/12/IJCT-Vol-2-Sabre-2017.pdf (Accessed: 6 July 2023).

Santesso, A. (2004) 'The birth of the birthplace: Bread Street and literary tourism before Stratford', *ELH*, 71 (2), pp. 377–403. Available at: https://www.jstor.org/stable/30030055 (Accessed: 11 August 2021).

Seaton, A.V. (1996) 'Guided by the dark: From thanatopsis to thanatourism', *International Journal of Heritage Studies*, 2 (4), pp. 234–244. Available at: https://doi.org/10.1080/13527259608722178 (Accessed: 10 March 2022).

Stone, P.R. (2006) 'A dark tourism spectrum: Towards a typology of death and macabre related tourist sites, attractions and exhibitions', *Tourism: An International Interdisciplinary Journal*, 54 (2), pp. 145–160. Available at: https://hrcak.srce.hr/161464 (Accessed: 13 August 2022).

Tilden, F. (2007) *Interpreting our heritage.* 4th edn. Chapel Hill: The University of North Carolina Press.

Trubek, A. (2011) *A skeptic's guide to writers' houses.* Philadelphia: University of Pennsylvania Press.

Wang, H.J. and Zhang, D. (2017) 'Comparing literary tourism in mainland China and Taiwan: The Lu Xun native place and the Lin Yutang house', *Tourism Management*, 59, pp. 234–253. Available at: https://doi.org/10.1016/j.tourman.2016.08.008 (Accessed: 13 August 2022).

Watson, N.J. (2006) *The literary tourist: Readers and places in Romantic and Victorian Britain*. Basingstoke: Palgrave MacMillan.

Watson, N.J. (2020) *The author's effects: On writer's house museums.* Oxford: Oxford University Press.

Westover, P. (2012) *Necromanticism: Traveling to meet the dead, 1750–1860*. Basingstoke: Palgrave MacMillan.

Woolf, V. (1980) *'Great men's houses' in The London scene: The essays* (first published 1932). London: Hogarth Press. pp. 19–24.

Young, L. (2015) 'Literature, museums, and national identity; Or, why are there so many writers' house museums in Britain?', *Museum History Journal*, 8 (2), pp. 229–246. Available at: https://doi.org/10.1179/1936981615Z.00000000052 (Accessed: 10 October 2021).

Young, L. (2019) 'Writers' house museums: English literature in the heart and on the ground', in McCracken Fletcher, L. (ed.) *Literary tourism and the British Isles: History imagination and the politics of the past*. Lanham, MD: Lexington Books, pp. 165–190.

Zemgulys, A. (2008) 'Henry James in a Victorian crowd: "The Birthplace" in context', *The Henry James Review*, 29 (3), pp. 245–256. doi:10.1353/hjr.0.0022.

5 Literary fiction places of the imagination

In the previous chapter we explored the many literary tourism places and spaces associated with the life of authors, whether they be homes, favourite haunts, or final resting places. We now move on to consider a form of literary tourism inspired by the fictional output of those writers who use place and its interaction with plot and characters to form their narrative. The resulting stories within their fictional locations are read, processed, and lived through the imaginations of the reader who may then be inspired to go out and seek the imagined landscapes within existing geographies. In their recent work on the role of imagination within popular culture, Reijnders *et al.* (2021, p. 9) remind us that although narratives may be seen as a collection of events, the location of this action is also fundamental as 'stories always literally take place somewhere.' This 'somewhere' may be a hazy fantasy land or a carefully described and recognisable city. The space wherein the narrative unfolds supports the story which 'leads to a situation in which many sites are known for the stories that take place in them, which may be symbolically associated at a later stage with the plot and moral themes of the story concerned' (Reijnders *et al.*, 2021, p. 9). This association can be very strong as Pocock suggested in his 1979 essay on 'The Novelist's Image of the North', in which he explores the many literary evocations of England's industrial northern regions, such as those of Elizabeth Gaskell, Benjamin Disraeli, Charles Dickens, the Brontës, Arnold Bennett, DH Lawrence, and George Orwell. The dark, polluted, harsh landscapes against which their characters battled, and through which the novelists made clear their social critique, provided an image of the North which continues to this day, despite the far-ranging social and economic changes which have taken place. Given the strength of this imagined landscape, 'the literary influence may well increase with time, and even persist when contrary evidence is available to refute the traditional image' (Pocock, 1979, p. 62). Despite the efforts of destination marketers to convey an alternative impression, that image of the North still pervades today.

In his study of literary tourism to sites associated with Dylan Thomas and Marcel Proust, Herbert (2001) comments that 'fiction may be set in locations that writers knew and there is a merging of the real and the imagined that gives such places a special meaning. Fictional characters and events often

DOI: 10.4324/9781003041740-5

generate the strongest imagery' (p. 314). Places associated with the fictional output of authors are therefore attractive, not just because they represent the settings depicted in the novels, although this can be compelling for readers, but for the additional interplay of favourite characters and dramatic events within those settings. We saw in Chapter Two that, historically, travel to seek discourse with authors and to meditate at their gravesides was undertaken by Grand Tourists and literary-inspired visitors but that this impetus shifted over time towards a greater engagement with the places and experiences associated with the text itself. This is first explained by the rise in Romanticism and the desire for emotion-infused, sublime landscapes and as the popularity of poets and poetry gave way to the rise of the realist novel in the nineteenth century, the range of available sites of literary tourism increased. Alfred, Lord Tennyson famously stated that his reason for visiting the popular and historic seaside resort of Lyme Regis was to see the setting of Jane Austen's *Persuasion* (1817), rather than any other attraction (Ousby, 1990) and he was not alone in this impetus, with his contemporaries searching the British Isles and beyond for the lakes, Highlands, mountains, waterfalls, and moors which they had first imaginatively visited through their reading of the works of Wordsworth, Scott, Byron, Rousseau, and the Brontë family, to name just a few. Today, the landscapes of fiction hold the same appeal, as the wide range of guidebooks and coffee table publications on the topic would suggest (see Box 1.2) in tandem with numerous television series which take viewers on journeys through the eyes of well-known and much-loved authors. For those less inclined to armchair travel, there is 'a thriving tourist industry presenting the place of the book for visitors, a sort of tourist anthology of the classics, canonical and popular of English Literature' (Watson, 2006, p. 2). A recent survey commissioned by Visit Nottinghamshire to uncover the levels of interest within the American market for British literary sites, discovered that these potential tourists were most likely to visit an English literary tourism destination based on its association with a real place and with literary characters. For these would-be literary tourists, visiting a place associated with the life of a writer was less of a draw (Ingram *et al.*, 2021).

In this chapter we will therefore explore a range of settings, both real and imagined, which have been used by authors in their works of fiction and which attract the literary fiction-inspired tourist. These places range from the purely fantastic to the topographically accurate and the chapter goes on to consider the nature of the visitor experience and the role of the reader-tourist in these locations. We discover that these settings exert a particular power over readers who visit them in order to bring their imaginative encounters with characters, places, and events to life. This form of concretisation of the imaginary can involve some work on the part of the literary tourist who may need to use the text itself to search for clues as not all literary settings lend themselves to incorporation within the tourism infrastructure. This co-creation between reader/visitor, place, text, and author can produce a potent literary tourism experience.

Lands of fantasy

Robinson (2002, pp. 51–54) categorises fictional places in three ways: 'fantastic lands,' 'invented and disguised places,' and 'writings and readings of "real" places and spaces.' Fantastic lands refer, of course, to places which have been wholly invented by the author and in which all the action takes place. Often these places are meticulously described, and the reader can easily inhabit them within their imaginations, without the regions having any stated connections to existing places. Famous examples such as Tolkien's 'Middle Earth,' C.S. Lewis' 'Narnia,' J.M. Barrie's 'Neverland,' and Lewis Carrol's 'Wonderland' show that these fictional countries are often found in literature aimed at younger readers. But other fantasy landscapes such as James Hilton's 'Shangri-la' and Jonathan Swift's 'Lilliput' and 'Blefuscu' have lured older readers over the centuries and their names and associations have entered our everyday language. Although completely fictional, these landscapes can encourage a desire to seek out the places that provided the original inspiration and tourism experiences may be created around this impetus. It is often noted, for example, that within Tolkien's 'Middle Earth' the Shire exudes Englishness and is often linked to the West Midlands, with its main town Hobbiton being on the same latitude as Oxford. Many regions of England claim the Shire as their own, including Hurst Green in the Ribble Valley, Lancashire where Tolkien worked on his trilogy whilst staying at Stonyhurst College where his son taught Classics. The local Tolkien Trail (one of many) claims that several names which appear in *The Lord of The Rings* trilogy (1954–1955) are similar to those found locally including Shire Lane (in Hurst Green) and the River Shirebourn (Visit Lancashire, 2022). James Hilton's post WW1 novel, *Lost Horizon* (1933) introduced the mythical, mountainous region of 'Shangri-la,' a fictitious paradise, full of long-lived, cultured peoples, sheltered from the outside world. The name became synonymous with an Edenic, lost innocence and entered the English language as a short cut for paradise, as a chain of up-market hotels, and as a popular and aspirational name for homeowners to bestow on their dwellings. Many attempts have been made to locate this utopian valley of lush forests, snow topped mountains, and lakes and in 2002, the county of Zhongdian in Yunnan province, China changed its name to 'Shangri-La' (Llamas and Belk, 2011). The region underwent a makeover to do justice to its new name with the construction of an international airport and large hotels complete with Buddhist iconography and 'infrastructure built to attract tourists also includes restaurants, shops, museums, a renovated Monastery, a reconstructed Old Town, a new temple, and paved roads leading to natural attractions' (Llamas and Belk, 2011, p. 258). The transformation here from fantasy land to topographical region is complete.

Places in disguise

Robinson's (2002, pp. 51–54) second category of fictional landscape, 'invented and disguised places', acknowledges the fact that literature is full of settings which are camouflaged versions of real places, or which may be composite regions, made up of different borrowings from several places with which the author was familiar and from which they might have drawn inspiration. The *Mapp and Lucia* novels of E.F. Benson (1920–1939) provide a good example of a thinly disguised real place – in reality, just a renaming. Benson's 'Tilling' is in fact the historic town of Rye in Sussex where Benson lived and wrote his novels in the same house that had previously sheltered Henry James. The name Tilling comes from the river Tillingham which flows through Rye and the E.F. Benson Society provide a walking tour map of the town showing which real sites stand in for fictional settings. Similarly, Somerset Maugham's fictional town of 'Blackstable' where the orphan Philip Carey comes to live with his aunt and uncle in *Of Human Bondage* (1915) is the Kent coastal town of Whitstable. Philip later goes to school in 'Tercanbury,' which is of course nearby Canterbury. Authors may feel the need to disguise their settings if they wish to be free to critique an area and its population; thus, Charles Dickens renamed the Lancashire town of Preston as 'Coketown' in *Hard Times*, (1854) his critique of the industrial system. Similarly, Thomas Hardy's bleak evocation of thwarted academic ambition in *Jude the Obscure* (1895) is set in 'Christminster' rather than Oxford and, more recently, Robin Jenkins' sometimes unflattering short stories of local life in the Scottish seaside town of 'Lunderston' hides the fact that he was using the people of Dunoon and its environs in Argyll as his inspiration for *Lunderston Tales* (1996).

Composite literary landscapes have been created by authors and used as settings for a number of books. Thomas Hardy famously did this with his creation of 'Wessex' (Dorset in the West of England) and in Mississippi, William Faulkner created the fictional county of 'Yoknapatawpha' which combines real and fictional features and provides the setting for *The Sound and the Fury* (1929) and other novels. Still in the United States of America (USA), Stephen King creates fictional towns within the real state of Maine and uses these as both settings for his novels and short stories and as references within them. His fictional town of 'Castle Rock' is the location for *The Dead Zone* (1979), *Cujo* (1981), and *The Body* (1982) and the town is mentioned in many others, thus creating a self-contained landscape within which the psychological horror can unfold, unabated. Agatha Christie used a composite approach when creating the village of 'St Mary Mead,' the fictional home of her detective Miss Marple and the setting of some of her stories. We are told that the village is within reach of London and is close to the sea, but it appears to be a conglomeration of peaceful village features from across the Home Counties. The imprecision, renaming, invention, and patchworking of literary settings described above means that these types of location do not easily lend themselves to local tourism strategies although the lack of geographical clarity is no barrier to some literary-inspired visitors hunting down clues in the search of favourite settings. As we saw in Chapter Two, in the case of Dracula fans, a lack of literary markers or references to vampires in the tourism promotion of Transylvania is no deterrent to their attempts to seek out the Count and track down likely locations (Light, 2017).

The real thing

Robinson (2002, pp. 51–54) describes his third type of fictional literary landscape as 'writings and reading of "real" places and spaces.' In this category of literary landscape, the author uses authentic settings as either a backdrop to events or in a more active way, where characters interact with the environment to progress the plot. Authors may vary in the level of detail that they offer and whether the setting is impressionistic or realised in a naturalistic manner. In some cases, the landscape or cityscape becomes a character within the story in its own right as we find with Egdon Heath in Thomas Hardy's *Return of the Native* (1878), the River Thames in Charles Dickens' *Our Mutual Friend* (1865), or the highway in Jack Kerouac's *On the Road* (1957). Often, the author will use places that are familiar to them to heighten the realism of their depiction, and this can result in novels that are so geographically accurate that they may be used as tour guides to a city or region. Accompanying maps might be produced that plot the movements of characters through the landscapes and this device is used in the development of tourist routes and trails and in guided tours as we shall see in Chapter Eight. Gustave Flaubert, considered to be one of the earliest exponents of literary realism in French literature, set his novel *Madame Bovary* (1856) in and around Rouen in Normandy, the town of his own birth. The region is described in meticulous and accurate detail as befits an author committed to the truthful representation of society and its environment and the Normandy tourism organisation offers a Flaubert Trail which includes biographical sites such as museums and statues of the author. The village of Ry, on which Flaubert based the fictional 'Yonville,' offers an In the Footsteps of Madame Bovary route with information panels and a digital app allowing visitors to replicate the fictional Emma Bovary's travels. Rather unusually, Ry's Museum of Automata is a gallery of 300 moving figures which represent the main scenes from the novel, thus further animating the literary landscape.

Another famous example of this type of topographical accuracy is James Joyce's depiction of Dublin in his novel *Ulysses* (1922) which follows the movements and encounters of Leopold Bloom and other characters, such as Stephen Dedalus, over the course of one day – 16 June 1904. Unlike *Madame Bovary, Ulysses* is very much a novel in the modernist tradition and as such it 'provokes readers to interpret the city spatially rather than viewing it as a narrative of sequential events' (Johnson, 2004, p. 92). This has allowed numerous maps and tourist guides to be produced which present Dublin through the lens of the novel, and which interpret the various routes taken by the characters, in particular the chief protagonist, Leopold, whose lunchtime journey from the *Evening Telegraph* office on Princes Street to the National Library on Kildare Street is represented in fourteen brass plaques inlaid in the pavement. Although highly impressionistic in style and using a stream of consciousness narrative, distinctive places in Dublin's geography clearly appear in the novel's episodic set pieces. One of these locations, the Martello

Tower at Sandycove, which appears at the opening of the novel, now houses the James Joyce Museum. It is often cited that Joyce intended *Ulysses* to be a faithful record of the city which could be used to recreate Dublin, should it ever be obliterated, and he certainly had ambitions in this novel, and in his collection of short stories *Dubliners* (1914), to give the city to the world (Bataillard, 2014, p. 39). The yearly Bloomsday celebrations held in Dublin and in other cities worldwide demonstrate an ambition realised as we shall see in Chapter Nine.

The examples above show literary associations that have built up over a long period of time, but a more recent fictional connection can ignite interest in an existing location, as has happened in the case of the mining heritage site of Denniston on the west coast of New Zealand's South Island. Denniston has become a popular attraction for fans of the historical fiction of author Jenny Pattrick, who set her two novels *The Denniston Rose* (2003) and *Heart of Coal* (2004) within this mining community which, at its peak, ran New Zealand's largest producing coal mine. She herself had visited the heritage site and was inspired by an interpretation panel describing the lives of the women in the nineteenth century township. The author then spent time at the site, talking to locals and understanding the geography before using the location and its history within her novels. Current interest in the site from readers has inspired a self-guided walking trail and a phone app that feature events and locations from the novels (New Zealand Tourism, 2022). This literary interpretation runs parallel with the site's designation as a Category One Historic Place with several relics on display and underground mining tours offered to industrial heritage tourists.

Some places have become so synonymous with their imaginative counterpart that the literary place name has replaced the topographical version. As we saw in Chapter Two, in Lake Katrine in the Trossachs lies the small island of Eilean Molach (Gaelic for 'the shingly island'). Now named 'Ellen's Isle' in both local parlance and on the ordnance survey map, this location reflects the importance of Walter Scott's long poem the *Lady of the Lake* (1810) and was the fictional home of 'the lady' Ellen Douglas. So many visitors came in search of the island described in the poem that the small, densely wooded isle became Ellen's own. It is also thought that the original Gaelic name, with 'eilean' meaning island, suggested the heroine's name 'Ellen' to Scott (Leask, 2020). Another example of place names altered to fit literary expectations can be found in the town of Illiers in the Centre-Loire Valley region of France which has close connections with the novelist Marcel Proust as we saw in Chapter Four. As a child, Proust visited his aunt and uncle at their home in Illiers and he used this experience in *Swann's Way* (1913), the first volume of his vast seven volume work *À la Recherche du Temps Perdu* (1913–1927). In *Swann's Way* he describes the house and the town, turning them into 'Aunt Léonie's house' and 'Combray,' respectively. This setting is key within Proust's work as it is in this house that the narrator first encounters the taste of the famous madeleine, the memory of which forms an epiphany for him later in the work. The blurring of the real and fictional worlds within the town became complete when Illiers renamed itself Illiers-Combray in 1974. Packets of madeleines (small, shell-shaped sponge cakes) of many different flavours are a popular souvenir for visitors to the region (Sabre, 2017) adding another sensory dimension to the experience of literary place.

The experience of fictional places

Having assessed the three different types of imaginative landscapes proposed by Robinson (2002) and available to literary tourists, in this section we explore the impetus that sets readers/ tourists off on the search of these fictional regions. What does the experience of such literary landscapes provide for the visitor and the reader and what do they seek in their travels? In considering the experience of literary and media tourists, Reijnders (2011) describes the search for the real locations of fictional settings as the pursuit of physical references to scenes and places that have first formulated in the imagination. He uses French historian Pierre Nora's theory of *lieux de memoire* (places of memory) to develop his theory of *lieux d'imagination* (places of the imagination). In Nora's case, *lieux de memoire* are places which anchor collective memory that accrues around, for example, national monuments or battlefields (Nora, 1989). These are settings which symbolise and validate national memory of events or historic periods. Reijnders believes that for literary and media tourism, this concept can also be applied to places of collective imagination – where real locations can fix in place the collective imaginings of those who have read a novel or watched a film. In Reijnders' research, these anchoring moments begin as soon as a reader realises that, for example, there really is a country called Transylvania, and a seaside town called Whitby and there might be a chance that they could travel in search of Count Dracula. This realisation 'is precisely where the transitional moment takes place, the instant when the world of the imagination coincides with – or possibly contrasts with – physical reality' (Reijnders, 2011, p. 234).

Reading the fictional landscape

Watson suggests that the type of interaction between reader, text, and place, discussed above, is what produces the idea of the literary landscape itself as 'literary place is produced by writing mediated by acts of readerly tourism, and in that sense literary place is itself a "text"' (Watson, 2006, p. 12). This viewpoint suggests that it is the novelistic representation of location, contemplated by the literary tourist through their visitation, which then produces a place which can be read as a literary artefact – a 'text.' And as a text, the literary place is open to many different interpretations and readings. Thus, the literary tourist will bring all manner of associations, prior knowledge, and preconceptions to their interaction with the literary landscape, whether they be specialists in the field of English Literature or long-suffering partners of literary tourists. Memories of childhood reading, current reading habits, exposure to destination management promotional strategies, or their consumption of televised or film adaptations of novels or literary travel guides will all colour the literary gaze to a greater or lesser extent. However, according to Booth (2016, p. 4), all visitors, 'whether expert or lightly informed about the commemorated writer, collaborate in extending the world of the text into physical space.'

As tourists, we are all influenced by the act of reading which prepares us for our travels, consciously or subconsciously. Laing and Frost (2012), in their text *Books and Travel*, suggest that fiction often uses travel itself as a key theme and plot device and that the popular trope of 'armchair travel' emphasises the power that fiction has to transport us to other times and places. Thus, literary visitors have already been primed by their reading and in his study of a group of visitors to Catherine Cookson Country in North-east England, Pocock suggested that these knowledgeable readers 'bring with them eyes already prepared to see' (Pocock, 1992, p. 40). For many readers, literary places first formulated within their imaginations will feel very real as they are able to people genuine places with imaginary characters and animate them with fictional events. This experience will then augment and enhance their (re)reading of the text. This process reminds us that 'to go to a place by the light of a book is at once to declare the place inadequately meaningful without the literary signification provided by the book, and to declare the book inadequate without this specific, anxiously located referent or paratext' (Watson 2006, p. 7). Thus, literary tourism highlights the symbiotic relationship between the act of reading and of literary touring – the geographical setting requires the fiction to come alive with romance and action and the novel needs the physical landscape in order to render it real. The text encourages the literary tour which then sends the reader back to the book, where a pressed leaf or gift-shop postcard now acts as bookmark and permanent reminder of the novel's setting, made real.

Realising the fictional landscape

In their study of self-confessed bibliophiles' experience of literary tourism, MacLeod, Shelley, and Morrison (2018) describe this process of extending the written text into geographical space as 'concretisation.' This happens when literary fans are motivated to make their reading experiences more tangible through acts of literary tourism. The theory of concretisation is aligned to reception theory in literary studies where the emphasis is on how the reader receives a text through the act of reading. The Polish philosopher Roman Ingarden was the first theorist to use the concept to explain the role that is played by a reader in realising a text as a literary work of art (Ingarden, 1973). No author can provide a complete universe or a fully rounded human character, although readers will experience their imaginings as such, due to their ability to 'flesh out the details and actualise the potential within the literary text. Indeed, it is the very gaps in the narrative that engage the reader as they work with the text to make meaning' (MacLeod et al., 2018, p. 391). This co-operation between the text and the reader/visitor is an act of co-creation where the visitor is no longer viewed as a passive consumer of goods and services but instead is an active producer of their own experiences (Campos *et al.*, 2015). Such visitors invariably engage in many types of literary tourism experience, but they do not always require a tourism infrastructure in order to pursue imaginative settings. We have already noted the intrepid literary fans seeking out Dracula in a region with little literary tourism promotion and, in the research carried out by MacLeod et al. (2018), bibliophile literary tourists used the texts and information from their own fan communities in order to plan their trips to fictional locations.

So far, we have considered the pull of literary landscapes as described in works of fiction and imaginatively reconstructed in the minds and in the touristic activities of visitors. However, in tandem with the literary locations and the undeniable draw of finding oneself in a landscape that hitherto one had only imagined, is the opportunity to encounter much-loved characters within these settings. In an oft-quoted example, Pocock (1987, quoted in Herbert, 2001) comments that, in his study, visitors to Haworth in Yorkshire are more inspired by the opportunity to inhabit the world of the rugged, tortured Heathcliff than the real-life milieu of the Brontë family. He suggests that, although visitors seek out the moors which inspired Emily to write *Wuthering Heights* (1847) and which are replete with references to the famous sisters, they are filled 'less with the excitement of treading in the Brontës' footsteps, than with the thought that Heathcliff might appear.' (p. 314). Similarly, visitors to Princc Edward Island, home of Lucy Maud Montgomery, are in search of her character Anne, of her *Anne of Green Gables* series, rather than the author herself, despite her clear biographical connections to the island (Gothie, 2016). Later in this chapter we will also see that the fictional character of detective Sherlock Holmes has become so thoroughly fleshed-out and inhabits his fictional landscape so fully that he has become more real than his more shadowy creator, Arthur Conan-Doyle.

Literary landscapes of childhood

Adult fans seek out the blue-remembered literary landscapes of their youth, undoubtedly motivated to quite an extent by a longing for the lost world of childhood reading. We saw in Chapter One that Squire (1994) considered this nostalgic element in an early study of literary tourism to Beatrix Potter locations in the Lake District where she found that adult visitors were using the sites to return to an idealised and comforting memory of childhood, as promoted by the familiar animal characters and settings. But visitors may also feel nostalgic for that period in childhood where, untrammelled by adult cares, one could become completely enveloped in the literary realm. Nicola Watson (2006), in observing her own offspring, concluded that children have no interest in literary tourism *per se*, although her daughters, and happily, the present author's daughter, will politely indulge their mothers in their forays. Children appear to be content to inhabit the imaginative landscapes of their reading without the need to make those lands tangible through visits. Thus, on a visit to the 'Hundred Acre Wood' in the Ashdown Forest in East Sussex where A.A. Milne and his son Christopher Robin had a country home and where he set his classic series of *Winnie the Pooh* books, the present author noticed that the experience was not of particular interest to a young accompanying visitor, although a game of Pooh Sticks on Pooh Sticks Bridge was enthusiastically received (see Fig. 5.1). Meanwhile, an adult pilgrim, approaching with her battered and much-loved bear in her arms was clearly greatly moved by the experience and by the opportunity to bring her Pooh back to his spiritual home. It appears that the golden lands of childhood reading, whether they be the fantasy universes of Middle Earth or real locations such as Susan Cooper's Buckinghamshire, Arthur Ransome's Lake District, or Laura Ingalls Wilder's prairie, will only be sought out when the child reader reaches nostalgic adulthood.

The place of biography in the fictional landscape

We have already seen that visitors can gain a potent experience through engagement with a fictional setting, whether it is to walk through the land of the imagination or chance upon a much-loved fictional character. However,

Figure 5.1 Pooh Sticks Bridge in the Ashdown Forest, Sussex. Photograph by the author.

Watson (2006) suggests that 'the experience is perhaps most powerfully compounded if the place of composition and fictional setting coincide' (p. 3) – in other words there is an optimum imaginative participation in places where biography and fiction collide. Examples of these types of sites include the Kim Yujeong literary village in the Chuncheon-si region of Korea. The famous short story writer Kim Yujeong (1908–1937) was born and lived in the village of Sille and produced works such as *Spring, Spring* (1935) and *Camellia Flower* (1936), all set within the village location. The popular works humorously depicted rural life in the 1930s and the village site includes Kim's recreated birthplace, exhibition halls, and statues of his fictional characters. A Story Road which is a trail around the village highlighting the fictional locations of Kim's stories has also been introduced (Lee and Weaver, 2014). Thus, the visitor can see the home of the author and the superimposed fictional setting of the stories within the same literary complex.

Orchard House, home of the Alcott family in Concord, Massachusetts offers a similar experience. We have already seen in Chapter Four that visitors can view Lucy May Alcott's desk where she wrote her much-loved novels and gain a *frisson* from some of the authentic artefacts on display such as her mood cushion which was also used by the author's alter ego, her fictional character, Jo March. Orchard House is the model for the March family's Concord home as described in the book series and in an analysis of Tripadvisor reviews of the house, MacLeod (2021) found that visitors often commented on the elision between place of composition and the imaginative setting, for example one stated that 'walking into the Alcott house is like walking into *Little Women* … you can almost see the girls practising plays and talking to each other' (p. 1220).

This coincidence of the biographical and the imaginary is clearly very potent, but authors can create compelling settings by using carefully researched real-life locations, even if they have no lived experience of them, as is the case with many of the plays of William Shakespeare. Much critical speculation has been undertaken as to whether the Bard travelled to his literary locations, in particular Italy, as he set nine of his plays there and they show an intimate knowledge of the places, customs, and climate. Some suggest that he could have gleaned this insight by talking to Italian merchants by the Thames in London whilst others point to the 'lost years' between 1585 and 1592 and ponder on whether he was in fact in Italy then (da Mosto, 2012). There is no evidence to suggest that he did travel overseas but this does not stop literary tourists from seeking out these locations, and destinations from making the most of their Shakespearean connections as Table 5.1 demonstrates. The location that has engendered the most touristic interest after Stratford-upon-Avon is 'fair Verona,' in Italy, the setting of *Romeo and Juliet*. Shakespeare based his play about these famous star-crossed lovers on an existing legend of warring families and the main attraction is Casa di Guilietta at 23 Via Capello which dates to the thirteenth century. Legend holds that this was the Capulets' house and some traces of their coat of arms can

Table 5.1 Shakespearian literary locations

Plays of William Shakespeare	*Setting and literary tourism experiences*
The Merchant of Venice 1596–1597	In this play, Venice in Italy is clearly the setting with characters referring to the Rialto district and to gondolas, both of which are still available for the contemporary visitor. Shakespeare placed his character Shylock in Venice's ghetto (the first in the world) and there is still a Jewish community resident in the city.
Hamlet 1599–1601	Shakespeare set this play about the troubled Prince Hamlet in Elsinore Castle which was inspired by Kronborg Castle in Helsingør in Denmark. The Castle has enthusiastically embraced its Shakespearean heritage with daily tours of literary locations and an annual Shakespeare festival.
Othello 1603–1604	Othello was set in both Venice and in 'a sea-port in Cyprus.' Visitors to Famagusta in Northern Cyprus can visit the fourteenth century castle standing at the entrance to the port. Originally called Harbour Citadel, the building was renamed Othello's Tower during the British colonial period (1878–1914) due to its location and alignment with Shakespeare's stage direction.
MacBeth 1606–1607	Shakespeare set 'the Scottish play' in a number of locations across Scotland which are now part of a MacBeth Driving Trail created by Visit Scotland. Several castles such as Glamis, Inverness, and Cawdor castles are included in the trail as is Moot Hill in Scone, the traditional coronation site of Scottish kings and where Shakespeare set MacBeth's crowning.

still be seen on the wall (Lipovšek and Kesić, 2014). Inside the house is a small museum and in the courtyard stands a bronze statue of Juliet. Visitors can access the balcony and enact the famous scene where the pair declare their love. The fact that Shakespeare does not mention a balcony in the play, and that this one was added to the house in 1936 to satisfy expectations, does not deter the (mainly teenaged) visitors who embody the play's characters through ritual activities such as rubbing the statue's breast for luck and posting up love letters onto the courtyard walls.

The city has become synonymous with romance and holds an annual 'Verona in Love' festival on 14 February whilst the site around Juliet's putative house attracts over a million visitors every year (Squires, 2022). Of course, the courtyard where Romeo and Juliet plight their troth is not the only key location in the play – the tomb of the doomed lovers is also presented to visitors. A sarcophagus, described rather dismissively as a 'water trough' by Dickens in his 1846 *Pictures from Italy* (quoted in Tessari, 2020, p. 304), lies empty, lidless, and uninscribed in the underground crypt of the church of San Francesco al Corso, an old Franciscan monastery in Verona. It has acquired an association with the historical antecedents of Shakespeare's characters and has long excited interest despite its lack of verified authenticity. Both Lord Byron and Napoleon

Bonaparte's wife, Marie-Louise, took away fragments of the tomb at their visits, he to give to his lover Augusta Leigh and she to make jewellery from the relics. When George Cukor made his Hollywood film of the play in 1936, the tomb was used in the set which has solidified its authenticity for visitors ever since.

Such ritualistic tourist practices remind us that the literary tour is a secular pilgrimage, but also point towards its performative and embodied nature. The co-creation of experience that takes place when visitors actively engage with sites and artefacts is clearly taking place here at Shakespeare's famous settings. Although many visitors will have read *Romeo and Juliet* at school, they will likely have had more exposure to performances of the play (whether in the theatre or on film) which will have demonstrated how to act out and embody the scene and thus take full possession of the literary landscape. As increasing numbers of young visitors take the opportunity to perform their own love story in Verona, concern about the continuing rise of tourist numbers at the Casa di Guilietta site has been mounting. The city council has attempted to introduce visitor management measures which have been voted out by local businesses who benefit from the tourist influx. At the time of writing, Verona's local authority is proposing to expropriate these local businesses in order to introduce turnstiles and online bookings to deal with overcrowding within the confined site (Squires, 2022).

Sleuths and the city

We have already considered the visitor experience of a wide range of literary locations, both rural and urban, and this section will end with a consideration of the importance of the city as literary setting. Urban settings have long been used by authors as a means of giving shape to their work and in using the shared experience of well-known cities, authors are found 'using street names and … the names of affluent or poor districts as cultural and economic markers which act as both a shorthand in their stories and as a device for greater verisimilitude' (Busby, Korstanje and Mansfield, 2011, p. 25). Famous cities have featured in literary works so often that they could in fact be said to have been created by the texts themselves within the imagination of visitors. We have already touched upon a number of urban settings in this chapter (for example Leopold Bloom's Dublin, Romeo and Juliet's Verona, and Proust's Illiers) but the specific focus here will be on the experience of the city as seen through the eyes of fictional detectives. Several researchers have paid particular attention to this specific genre of literary tourism, for example van Es and Reijnders (2016, 2018), Reijnders, (2009), Effron, (2021), and McLaughlin, (2016) who explore the London of Sherlock Holmes, Philip Marlowe's Los Angeles, Lisbeth Salander's Stockholm, and the Oxford of Inspector Morse. van Es and Reijnders (2016) suggest that the popularity of crime fiction and the subsequent tours, trails, and walking guides which have been produced to allow crime fans to explore the city through a detecting lens, were first a backlash against the rural, Romantic picturesque which itself was an antidote and escape from the burgeoning urban sprawl of the nineteenth century. Much literary output (and literary tourism) still attends to the Romantic and the rural but there is also a strong fascination for the darker aspects of city life, whether that be crime, corruption, or poverty. The specific lure of the urban detective (for the reader and the tourist) can be explained by the need to both face up to the realities of city life and contain them. Thus, exploring the city through the eyes of Holmes, Marlow, or Salander is to satisfy the 'cultural desire of seeing the ails and mysteries of urban life exposed, thoroughly analysed and ultimately controlled through the supreme observational and reasoning skills of the main protagonist(s)' (van Es and Reijnders, 2016, p. 116).

Effron (2021) discusses the ways in which the *Mystery Readers' Walking Tour* guidebooks have influenced the reading of cities. These guides to London (1987), New York (1993), Chicago (1995), and Washington (1998) reconfigure the urban spaces as scenes of mystery stories, using traditional guidebook conventions and mapping to illuminate purely fictional happenings – in effect, 'reshaping reality through mystery fiction structures' (Effron, 2021, p. 1373). This allows the reader and visitor to descend deeper into the dark underbelly of the city through their reading and their tourist practice, engaging with the city in a more profound way (van Es and Reijnders, 2016). This descent may be considered an aspect of dark tourism (see Chapter Four) where visitors are fascinated by death and its trappings, but the depth of engagement might also address concerns about the placeless and superficial nature of globalised urban life. The fan of crime and mystery fiction is not motivated purely by the darkness of the themes but by the mystery and its detection which unfolds on the city streets. Therefore, fans of Sherlock Holmes are 'looking for, rather than simply looking at, actual-world locations that belong in Holmes' world' (McLaughlin, 2016, p. 146) and the reader-tourist consequently becomes the embodied detective, following footsteps and searching for clues.

We have seen in Chapter One that the global Covid-19 pandemic boosted reading of all types and formats, and this includes crime and mystery fiction which is one of the most popular genres worldwide. In the United Kingdom (UK), crime is the bestselling genre with book sales continuing to rise (Flood, 2022). Mystery stories such as Richard Osman's *The Thursday Murder Club* (2020) and its sequel *The Man Who Died Twice* (2021) were both in the top five best sellers in the UK in 2021. With their appeal for wide audiences, tales of detection and murder mystery have long been considered more populist than the canonical works of fiction which form the basis of much established literary tourism. van Es and Reijnders (2016) suggest that traditional forms of literary tourism are akin to heritage tourism in their pursuit of long-deceased authors and historical settings. The conventional literary tourist wishes to commune with the artist and the settings in meditative isolation, applying the Romantic gaze. Crime fiction on the other hand appeals to the contemporary literary tourist who is often taking part in a guided literary tour and enjoying a more collective gaze upon the dark heart of the modern metropolis (van Es and Reijnders, 2018).

However, despite the popularity of contemporary crime writers such as Stieg Larsson, Ian Rankin, and Colin Dexter and their associated tours of Stockholm, Edinburgh, and Oxford, the time-honoured sleuths from the pen of Arthur Conan Doyle and Agatha Christie still loom large. Christie, whose work does not often engage with the gritty metropolis, created perennial detectives, Hercule Poirot, Miss Marple, Tommy and Tuppence, and Parker Pyne and is celebrated in 'Agatha Christie Country' in Devon, an assimilation of biographical and literary locations. Here, an Agatha Christie Festival is held annually in Torquay, and visitors can follow a self-guided walking trail, or promenade on the Agatha Christie Mile. Her nearby former holiday home, Greenway, opened to the public in 2009 and locations such as the Burgh Island Hotel can be visited (Eichhammer, 2016). Arthur Conan Doyle's Sherlock Holmes, still the archetypal fictional detective, is so thoroughly embedded within the collective imagination through constant retellings and film and television adaptations, that he has an existence that seems far more enduring and manifest than that of his creator. Despite being a fictional character, Holmes boasts not one but two addresses – the famous Baker Street abode in London and a small flint cottage in the village of East Dean in Sussex where a blue plaque tells us that 'Sherlock Holmes, consulting detective and beekeeper retired here 1903–1917.' Conan Doyle, on the other hand, leaves little trace for such a celebrated author. Even his place of birth in 1859 in Picardy Place, Edinburgh, is marked by a statue of his fictional creation, contemplating his pipe. Further statues of Holmes can be found outside 221b Baker Street, near the Reichenbach Falls in Meiringen, Switzerland, in Karuizawa, Japan, and, with Dr Watson, outside the British Embassy in Moscow. Conversely, a small statue of Conan Doyle is tucked away in the village of Crowborough, Sussex where the author spent his last years and there is a blue plaque on his former home in South Norwood, South London.

Interpreting the fictional landscape

As has become apparent from the discussion above, these different types of fictional setting lend themselves to interpretation with varying degrees of ease and success. Curators of sites of biographical interest have various choices as to the presentation and display of, for example, authors' homes, which may be set out in a naturalistic manner with minimum interpretation to promote a sense of domesticity and authorial proximity, or which might resemble a traditional museum with glass cases, interpretation panel, and a more didactic approach (see Chapter Four). These sites are generally bounded, which allows for a more focused interpretation by site managers who have control of the message being promoted. Imaginative landscapes of fiction can pose far greater challenges for those who wish to promote and explain their literary links to a potential touristic audience. In some cases, as we have seen, an author has created a wholly fantastic world which is difficult to anchor to any geographical reality although both fans and destination management organisations have tried, as we saw in the example above of Tolkien's Shire. Some imaginative landscapes may even be contested as we see in Box 5.1 which explores the identification of the setting of George Orwell's *Animal Farm* (1945).

Box 5.1 Case study: On the farm with George Orwell

George Orwell's famous critique of totalitarianism, *Animal Farm: A Fairy Story*, was published in 1945 and since then has never been out of print. In this novel, Orwell explores what might happen when the farm animals drive the humans out and take power over for themselves. The farm in the tale, Manor Farm in the village of Willingdon, becomes Animal Farm and much of the action takes place in the Great Barn where Old Major, the wise middle white boar inspires the other animals to believe that a new order may be possible where humans are no longer in power, and the animals can enjoy the fruits of their own labour. The plan does not work out as hoped, however, and the animals' lives do not improve. Orwell intended that his fable would represent the events leading up to the Russian Revolution of 1917 and the subsequent Stalinist era of the Soviet Union.

Orwell was living in the North Hertfordshire village of Wallington in the years before he wrote *Animal Farm* (1936–1940), and just across from the cottage named The Stores, that he occupied with his wife Eileen, lay Manor Farm, the Great Barn, and its herd of pigs which became the inspiration for the literary work. Other locations have claimed the Orwellian connection; for example, the owners of Chalk Farm in the East Sussex village of Willingdon, who run a charity that provides work experience for young adults with learning disabilities, were keen to find a connection to help them unlock English Heritage funding for roof repairs. Along with the compelling clue in the village name and similarities to the farm described in Orwell's tale, the site was also close to St Cyprian's School in Eastbourne where Orwell was a pupil (Argus, 1999). However, it seems that all the evidence now points towards the Hertfordshire site as being the potential model of the literary setting. Orwell and his wife set up a village store in their cottage on Kits Lane and had a smallholding where they grew fruit and vegetables, and raised chickens and goats, keeping a diary outlining the health of their animals and their crops. The goats were named Muriel and Kate, with Muriel later finding her way into the novel. Orwell's experience in caring for the animals is reflected in the characteristics that he gives to his protagonists in *Animal Farm* – the animals are given appropriate traits and are fully rounded individuals (Country Life, 2020).

The cottage and the farm are easily visible to the literary tourist who wishes to make a pilgrimage to the author's home and setting of this famous work but there is little in the way of interpretation. A small red plaque affixed to the cottage by Hertfordshire County Council in 1989 states that 'George Orwell (1903–1950), author lived here (1936–1940)' and there is a small exhibition on Orwell's life in Wallington in the local church where Orwell and his wife Eileen O'Shaughnessy were married. Hertfordshire's tourism organisation, Visit Herts, includes the site in their 'Celebrate Hertfordshire's Literary Links' webpage (Visit Herts, 2022).

Despite Orwell's fame and stature as an author, there was no museum dedicated to him until very recently. His birthplace in Motihari in the state of Bihar in India (where he lived for one year) has just opened as a museum (Rahman, 2014) and, in 2022, the Museum of Wigan Life opened a permanent collection of material related to the author's life and works, in particular those pertaining to his sociological study, *The Road to Wigan Pier* (1937). In 2017, a bronze statue of the author was placed outside the British Broadcasting Corporation building in London where he worked from 1941–1943. Barnhill Farm on the island of Jura where he wrote *Nineteen Eighty-Four* (1949) is available to rent as holiday accommodation as we saw in Table 4.1.

Fictional landscapes can cover wide regions and encompass several sites and businesses which may or may not be tourism-related, making it difficult to create and maintain a unified interpretation strategy. Literary associations may be outlined in the webpages of the local destination management organisation and specific sites related to the fictional action can be mapped onto existing cartography. Information panels may be provided at gateway points leading into such literary landscapes, providing details on texts, scenes, characters, and authors and their associated locations. As we saw in Chapter Two, authors and their fictional creations which have stood the test of time may even be accorded a 'country' as a means of identifying a region of both fictional and biographical interest. Road signs indicating that you are now in literary territory demarcate these fictional and biographical landscapes. Examples of these include 'George Eliot Country' in the Midlands of England and 'Mark Twain Country,' Elmira, New York. The promotional power of these literary countries which follow no administrative or political boundaries but instead reflect the geographical contours of the author and their work has long been understood (Robinson and Andersen, 2002; Hoppen, Brown and Fyall, 2014) and will be further examined in Chapter Seven where we investigate places of literary reputation.

Guided tours with a knowledgeable guide will pinpoint the sites of fictional action for literary tourists with the added benefit of like-minded participants with whom to share the experience. This *communitas* of pilgrims may participate further with a series of ritualistic activities such as dressing up in appropriate costume and reading or performing extracts from the work *in situ*, thus continuing a long tradition first developed during the Grand Tour where visitors sought to read out favourite lines in appropriate locations (see Chapter Two). The performative aspects of this type of engagement will be considered further in Chapter Nine. Self-guided trails may simply offer highlights marked on an existing map, such as the map of 'Tilling' (Rye in East Sussex) that explores the locations of E.F. Benson's tales of *Mapp and Lucia*. More complex audio or digital tours use mobile devices to map out and interpret a themed journey relating to the movements of a fictional character. Those movements may be leisurely and local, like those of Virginia Woolf's *Mrs Dalloway* (1925) and her contemplative wander through London, or significantly more expansive and energetic as in the case of Cervantes' Don Quixote and Sancho Panza's 2,500 kilometres themed route through Spain's Castilla-La Mancha, based on the two-volume novel, *Don Quixote* (1605 and 1615). Literary routes and trails will be considered in more detail in Chapter Eight. Literary events and festivals, one of the subjects of Chapter Seven, can also animate and interpret fictional landscapes as we have already noted in the example of James Joyce's Dublin and in the Shakespeare Festival held at Kronborg Castle in Helsingør, Denmark.

Conclusion

In this chapter we have explored the geographical realisation of the imaginary landscapes found in works of fiction and considered why literary visitors would be drawn to these settings and the nature of the experience they may offer. In this exploration we have considered the three types of fictional landscapes outlined by Robinson (2002, pp. 51–54), namely 'fantastic lands,' 'invented and disguised places,' and 'writings and readings of "real" places and spaces.' Lands of fantasy have been created wholly within the author's mind and these imaginary spaces are often very real to readers who inhabit them as they read. It is challenging to anchor these lands within the real world although, as we have seen, attempts are often made by literary tourists and destination management organisations to try to locate the geographical inspiration for these imaginary places. 'Invented and disguised places' refer to literary settings which are real places which have been camouflaged by the author (perhaps by a renaming) or which have been created by bringing together a patchwork of features from real places to create a new literary landscape. These lands are perhaps easier to track down than fantastical regions and, in some cases, the inspiration for the disguised setting is well known and can be promoted by the destination. However, this is not always true and, in these situations, literary tourists keen to experience the settings *in situ* need a good knowledge of the texts and a willingness to undertake some detective work in tracking down clues. In Robinson's (2002) third category, '"real" places and spaces,' the author uses real-world settings against which the action unfolds. Sometimes this authentic setting is very accurately described and could be used as a guide to potential literary tourists. In other cases, the setting is genuine, but its realisation is hazier. These imaginative settings allow the visitor to actively engage with places, bringing their literary encounters to life. In inhabiting, embodying, and in some cases co-creating the literary landscape, the visitor can bring new dimensions to their consumption of literary place and to bring that experience back to their reading of favourite texts. In the next chapter, we move away from locations associated with the biography and imagination of the author to explore the case of socially-constructed sites where literary themes are exploited to create new attractions or extend existing ones.

References

Argus (1999) 'By George, has the animal farmland been found?', *The Argus*, 19 May. Available at: https://www.theargus.co.uk/news/6799597.by-george-has-the-animal-farm-land-been-found/ (Accessed: 8 July 2023).

Bataillard, P. (2014) 'Musing in and out of the museum with James Joyce's Ulysses', *Word & Image*, 30 (1), pp. 39–45. Available at: https://doi.org/10.1080/02666286.2013.771930 (Accessed: 12 June 2022).

Booth, A. (2016) *Homes and haunts: Touring writers' shrines and countries.* Oxford: Oxford University Press.

Busby, G, Korstanje, M. and Mansfield, C. (2011) 'Madrid: Literary fiction and the imaginary urban destination', *Journal of Tourism Consumption and Practice*, 3 (2), pp. 20–37. Available at: http://hdl.handle.net/10026.1/11577 (Accessed: 14 June 2022).

Campos, A.C., Mendes, J., Valle, P.O.D. and Scott, N. (2015) 'Co-creation of tourist experiences: A literature review', *Current Issues in Tourism*, 21 (4), pp. 369–400. Available at: https://doi.org/10.1080/13683500.2015.1081158 (Accessed: 3 November 2022).

Country Life (2020) 'The real-life farm and animals that give life to George Orwell's Animal Farm', *Country Life*, 28 June. Available at: https://www.countrylife.co.uk/out-and-about/the-real-life-farm-and-animals-that-george-orwells-animal-farm-the-real-animals-and-fa-216491 (Accessed: 8 July 2023).

da Mosto, F. (2012) 'Shakespeare's Italy', *BBC Travel*. Available at: https://www.bbc.com/travel/article/20120416-shakespeares-italy (Accessed: 18 October 2022).

Effron, M. (2021) 'Reshaping reality: Mystery fiction literary tourism and its effect on real-world spaces', *The Journal of Popular Culture*, 54 (6), pp. 1371–1393. Available at: https://doi.org/10.1111/jpcu.13087 (Accessed: 20 October 2022).

Eichhammer, B. (2016) 'Topographies of detection: Literary tourism, English heritage and the making of Agatha Christie Country', in Habermann, I. and Keller, D. (eds.) *English topographies in literature and culture: Space, place, and identity.* Leiden: Brill, pp. 134–157.

Flood, A. (2022) 'UK book sales in 2021 highest in a decade', *The Guardian*, 11 January. Available at: https://www.theguardian.com/books/2022/jan/11/uk-book-sales-in-2021-highest-in-a-decade (Accessed: 8 July 2023).

Gothie, S.C. (2016) 'Playing "Anne": Red braids, green gables, and literary tourists on Prince Edward Island', *Tourist Studies*, 16 (4), pp. 405–421. doi:10.1177/1468797615618092.

Herbert, D. (2001) 'Literary places, tourism and the heritage experience', *Annals of Tourism Research*, 28 (2), pp. 312–333. Available at: https://doi.org/10.1016/S0160-7383(00)00048–00047 (Accessed: 24 October 2022).

Hoppen, A., Brown, L. and Fyall, A. (2014) 'Literary tourism: Opportunities and challenges for the marketing and branding of destinations?', *Journal of Destination Marketing & Management*, 3 (1), pp. 37–47. Available at: https://doi.org/10.1016/j.jdmm.2013.12.009 (Accessed: 26 October 2022).

Ingarden, R. (1973) *The literary work of art*. Translated by G.G. Grabowicz. Champaign, Il: University of Illinois.

Ingram, C., Themistocleous, C., Rickly, J.M. and McCabe, S. (2021) 'Marketing "Literary England" beyond the special interest tourist', *Annals of Tourism Research Empirical Insights*, 2 (2), pp. 1–10. Available at: https://doi.org/10.1016/j.annale.2021.100018 (Accessed: 7 July 2023).

Johnson, N.C. (2004) 'Fictional journeys: paper landscapes, tourist trails and Dublin's literary texts', *Social & Cultural Geography*, 5 (1), pp. 91–107. Available at: https://doi.org/10.1080/1464936042000181335 (Accessed: 7 November 2022).

Laing, J. and Frost, W. (2012) *Books and travel*. Bristol: Channel View Publications.

Leask, N. (2020) *Stepping westward: Writing the highland tour c. 1720–1830*. Oxford: Oxford University Press.

Lee, Y. and Weaver, D. (2014) 'The tourism area life cycle in Kim Yujeong literary village, Korea', *Asia Pacific Journal of Tourism Research*, 19 (2), pp. 181–198. Available at: https://doi.org/10.1080/10941665.2012.735681 (Accessed: 10 November 2022).

Light, D. (2017) 'The undead and dark tourism: Dracula tourism in Romania' in Hooper, G. and Lennon, J.J. (eds.) *Dark tourism: Practice and interpretation*. Abingdon: Routledge, pp. 121–133.

Lipovšek, E. and Kesić, S. (2014) 'Journeys beyond pages: The use of fiction in tourism', *Turističko poslovanje*, 14, pp. 51–60. Available at: https://doi.org/10.5937/TurPos1414051L (Accessed: 22 November 2022).

Llamas, R. and Belk, R. (2011) 'Shangri-La: Messing with a myth', *Journal of Macromarketing*, 31 (3), pp. 257–275. doi:10.1177/0276146711406549.

MacLeod, N. (2021) '"A faint whiff of cigar": The literary tourist's experience of visiting writers' homes', *Current Issues in Tourism*, 24 (9), pp. 1211–1226. Available at: https://doi.org/10.1080/13683500.2020.1765996 (Accessed: 2 December 2022).

MacLeod, N., Shelley, J. and Morrison, A.M. (2018) 'The touring reader: Understanding the bibliophile's experience of literary tourism', *Tourism Management*, 67, pp. 388–398. Available at: https://doi.org/10.1016/j.tourman.2018.02.006 (Accessed: 2 December 2022).

McLaughlin, D. (2016) 'The game's afoot: Walking as practice in Sherlockian literary geographies', *Literary Geographies*, 2 (2), pp. 144–163. Available at: https://doi.org/10.17863/CAM.9474 (Accessed: 3 December 2022).

New Zealand Tourism (2022) Denniston literary trail. Available at: https://www.newzealand.com/nz/feature/denniston-literary-trail/ (Accessed: 8 July 2022).

Nora, P. (1989) 'Between memory and history: Les lieux de mémoire', *Representations*, 26, pp. 7–24. Available at: https://doi.org/10.2307/2928520 (Accessed: 7 July 2023).

Ousby, I. (1990) *The Englishman's England. Taste, travel and the rise of tourism*, Cambridge: Cambridge University Press.

Pocock, D. (1992) 'Catherine Cookson country: Tourist expectation and experience', *Geography*, 77 (3), pp. 236–243. Available at: https://www.jstor.org/stable/40572194 (Accessed: 3 April 2021).

Pocock, D.C. (1979) 'The novelist's image of the North', *Transactions of the Institute of British Geographers*, pp. 62–76. Available at: https://doi.org/10.2307/621924 (Accessed: 3 March 2021).

Rahman, M. (2014) 'George Orwell's birthplace in India set to become a museum', *The Guardian*, 30 July. Available at: https://www.theguardian.com/world/2014/jun/30/george-orwell-birthplace-motihari-bihar-india-museum (Accessed: 12 December 2022).

Reijnders, S. (2009) 'Watching the detectives: Inside the guilty landscapes of Inspector Morse, Baantjer and Wallander', *European Journal of Communication*, 24 (2), pp. 165–181. doi:10.1177/0267323108101830.

Reijnders, S. (2011) 'Stalking the count: Dracula, fandom and tourism', *Annals of Tourism Research*, 38 (1), pp. 231–248. Available at: https://doi.org/10.1016/j.annals.2010.08.006 (Accessed: 14 December 2022).

Reijnders, S., Waysdorf, A., Bolderman, L. and van Es, N. (2021) 'Introduction', in van Es, N., Reijnders, S., Bolderman, L. and Waysdorf, A. (eds.) *Locating imagination in popular culture: Place, tourism and belonging*. Abingdon: Routledge, pp. 1–16.

Robinson, M. (2002) 'Between and beyond the pages: Literature-tourism relationships', in Robinson, M. and Andersen, H.C. (eds.) *Literature and tourism: Reading and writing tourism texts*. London: Continuum, pp. 39–79.

Robinson, M. and Andersen, H.C. (2002) 'Reading between the lines: Literature and the creation of touristic spaces', in Robinson, M. and Andersen, H.C. (eds.) *Literature and tourism: Reading and writing tourism texts*. London: Continuum, pp. 1–38.

Sabre, C. (2017) 'Between literary tourism and contents tourism: Marcel Proust and his Madeleine as a tourist attraction', *International Journal of Contents Tourism*, 2, pp. 1–11. Available at: https://contents-tourism.press/wp-content/uploads/2017/12/IJCT-Vol-2-Sabre-2017.pdf (Accessed: 12 December 2022).

Squire, S.J. (1994) 'The cultural values of literary tourism', *Annals of Tourism Research*, 21 (1), pp. 103–120. Available at: https://doi.org/10.1016/0160-7383(94)90007-90008 (Accessed: 20 April 2021).

Squires, N. (2022) 'Row erupts as tourists queuing to rub Verona's famous Juliet statue force council's hand', *The Telegraph*, 13 May. Available at: https://www.telegraph.co.uk/world-news/2022/05/13/row-erupts-tourists-queuing-rub-famous-juliet-statue-force-councils/ (Accessed: 25 May 2022).

Tessari, A. (2020) 'Verona and the myth of Romeo and Juliet: Love affair or merely affairs?', *Journal of Tourism History*, 12 (3), pp. 298–316. Available at: https://doi.org/10.1080/1755182X.2020.1830189 (Accessed: 26 May 2022).

van Es, N. and Reijnders, S. (2016) 'Chasing sleuths and unravelling the metropolis: Analyzing the tourist experience of Sherlock Holmes' London, Philip Marlowe's Los Angeles and Lisbeth Salander's Stockholm', *Annals of Tourism Research*, 57, pp. 113–125. Available at: https://doi.org/10.1016/j.annals.2015.11.017 (Accessed: 25 May 2022).

van Es, N. and Reijnders, S. (2018) 'Making sense of capital crime cities: Getting underneath the urban facade on crime-detective fiction tours', *European Journal of Cultural Studies*, 21 (4), pp. 502–520. Available at: https://doi.org/10.1177/1367549416656855 (Accessed: 25 May 2022).

Visit Herts (2022) Celebrate Hertfordshire's literary links. Available at: https://www.visitherts.co.uk/visit-herts-blog/celebrate-hertfordshires-literary-links/ (Accessed: 4 June 2022).

Visit Lancashire (2022) The Tolkien Trail – In the footsteps of J.R.R. Tolkien. Available at: https://www.visitlancashire.com/things-to-do/the-tolkien-trail-in-the-footsteps-of-j-r-r-tolkien-p583810 (Accessed: 25 May 2022).

Watson, N.J. (2006) *The literary tourist: Readers and places in Romantic and Victorian Britain*. Basingstoke: MacMillan Palgrave.

6 The social construction of literary places

In this book we have so far considered literary places associated with an author's life, output, and death and settings which attract attention from visitors because they have featured in works of fiction. In both cases, the authenticity of the biographical site and the verisimilitude of the imaginary setting with the real-world location are considered to be central to the power of these places and their ability to attract literary visitors. However, not all literary tourism sites fall into these two categories, and we will now go on to examine what have been termed 'socially-constructed' literary places – sites which have been deliberately created around a literary theme. Herbert (2001, p. 313) comments on the purposeful nature of such attractions in contrast to the inadvertent characteristics of biographical sites, thus: 'Literary places are no longer accidents of history, sites of a writer's birth or death; they are also social constructions, created, amplified, and promoted to attract visitors.' In this chapter we will explore the ways in which this deliberate creation, amplification, and promotion of a literary theme takes place and how it is experienced by the visitor.

In Chapter Three, we considered the commodifying potential of tourism as part of our wider consumer culture. Various critiques have been levelled at the impact that commercialised travel can have on the host communities, their culture, and lifestyles and on the experience of the visitor themselves. In particular, the process of commodification is seen to take its toll on the essential authenticity that places and culture are thought to possess (see Chapter Three for a fuller discussion on the concept of authenticity). The sites and experiences associated with literary tourism have not been immune to this censure. Courtney Watson (2013), for example, discusses the 'commodification of the South' which has used the work and biographical detail of writers such as William Faulkner to produce a spurious authenticity which she believes limits the way that visitors can experience literary tourism sites in Mississippi. Lipovšek and Kesić (2015, p. 106) refer to the 'disparaging and banal destination image' focussed purely on commercial gain that can be the result of a literary re-imagining of place. In his study of literary tourism in Sweden, Müller (2006) notes that such commodification of the literary theme brings additional planning responsibilities for regions wherein this literary

DOI: 10.4324/9781003041740-6

exploitation takes place, both in terms of managing increased visitor numbers and incorporating the views of the host community.

The commodification process goes through a number of stages as Fox, (2008, pp. 144–145) outlines in relation to presenting the relatively unknown Croatian playwright and poet Marin Držić (1508–1567) to the public. These stages involve the *vision* stage where destination managers first become aware of the potential of the literary figure and the *explication* stage where the literary value is translated into concepts that can be utilised within the tourism destination. Next comes the *conversion* stage where aspects of the author's life and work are tailored to suit the needs and tastes of the intended audience. Finally, the *benefits* stage underlines the positive tourism impacts resulting from the commodification process. Critics may assert that the essential meaning of a writer's work is lost through this practice of selection and presentation although, as Robinson and Andersen (2002) point out, the interpretation of a literary work is open to many different readings, thus making the search for authenticity within a literary tourism context irrelevant. Whilst we may be able to visit the actual birthplace, grave, or home of an author, accessing a genuine interpretation of their literary output may be impossible. In trying to determine what the literary visitor is really looking for in travelling to these sites, it is important to remember the role of the imagination which has allowed them to 'pre-visit' literary landscapes before arriving bodily at a location. Thus 'the idea of place formed in the literary tourist's mind may seem more authentic than the actual site once visited' (Stiebel, 2004, p. 34) and a visitor travels merely to reinforce the perceptions that they already hold (Watson, 2013). The concept of commodification is inherent in all aspects of literary tourism, but it is perhaps most closely aligned to the types of sites that we are examining here – socially-constructed sites that have deliberately used a literary persona or themes to create a visitor attraction *ab initio* or to extend the potential of an existing site.

Although all socially-constructed sites are purposefully created in order to produce an attraction for commercial purposes, for regional development, or educational rationales, it is possible to identify two distinct types of sites within this category. The first is an attraction in which an existing literary association is extended and exploited, for example the creation of a visitor centre at an established literary location which offers tourists an immersive and participatory form of interpretation, augmenting the literary link. For the purposes of this chapter, these types of socially-constructed literary site will be referred to as attractions which *extend* the literary theme. In some cases, these attractions have been created directly from the imaginary literary landscape and are what Jean Baudrillard (1983) termed *simulacra* – a copy of something that has never existed. The second type of socially-constructed literary site is more free-floating, providing a decontextualised experience that is founded on a recognisable literary motif. This might involve building replicas of existing literary attractions such as the recreated Shakespearean sites found in China and Japan (see Box 2.1 and Box 6.1). And as we shall see in this chapter, numerous examples of this type of attraction use the archetypal trope of the fairy tale to develop themed attractions loosely based on children's literature and aimed at families. In the discussion below, these sites will be referred to as places which *produce* the literary theme.

Extending the literary theme

Literary sites which purposefully build on an existing biographical setting or recognisable fictional location can be seen to be extending the literary theme. This may be done by curators or managers of existing sites to offer a more complete experience to visitors, or by employing innovative interpretation strategies to create a new attraction which builds on the extant literary link. One example of such a site is Casa Museo Mario Vargas Llosa in Arequipa, Peru which is the birthplace and early home of the Nobel Prize winning novelist. At this site, technology has been employed to create an immersive visitor experience within what is, at first sight, a traditional writer's house museum. Mario Vargas Llosa was born in the house at 101 Calle Parra in 1936 and lived there for his first year with his mother and her family, before moving to Bolivia. The regional government of Arequipa founded the museum in 2014 to honour Peru's first Nobel Prize winner. The author himself was involved in setting up the museum, donating books, portraits, and personal belongings, as was his cousin, the filmmaker Lucho Llosa, whose influence can be detected in the interpretive strategy employed.

The large house has 16 rooms displaying different aspects of the author's life, incorporating traditional settings with furniture, books, and paintings. However, Vargas Llosa himself appears throughout the tour as a hologram, welcoming the visitor to his home. This technique is dramatically extended into the biographical realm with a hologram of his mother giving birth to him in one of the rooms. He also appears in conversation with characters from his novels and settings from his work are recreated, described in the *Lonely Planet Guide to Peru* as 'rooms-turned-stages of scenes from his novels' (Lonely Planet, 2022). The house bore witness to the first year of the author's life, but it does not shy away from interpreting the rest of his decades spent in South America and Europe. There are exhibitions relating to his time in London and Barcelona and recreations of his room in Barranco in Lima and his attic in Paris can be visited. The house can only be accessed through a guided tour and visitors arriving on March 28th might catch Vargas Llosa himself on his annual birthday visit.

Moat Brae, incorporating the new National Centre for Children's Literature and Storytelling in Dumfries, Scotland opened in 2019 and is another example of a socially-constructed site that extends an existing literary link. In this particular case, the justification for developing the site was educational, with partners including National Lottery Heritage Fund, the Scottish government, and Dumfries and Galloway council, amongst others. Run by the Moat Brae Trust, the centre was set up to be both a place of learning and an international visitor attraction. Moat Brae is the former home of author J.M. Barrie's childhood friends Stuart and Hal Gordon and he himself attended Dumfries Academy which is situated nearby. He was a frequent visitor to the large house and reported that the garden was the 'enchanted land' which inspired him to write his most famous work, the play *Peter Pan*, first produced in 1904 (Brooks, 2019). The house and gardens have been restored, the Neverland Discovery Garden has a pirate ship, crocodiles, and mermaids for visitors to explore and within the house an interactive experience has been created. One of the rooms is a recreation of the Darling children's nursery where visitors can chase Peter Pan's shadow before crawling through Nana, the dog's, kennel and into Neverland itself. A playroom full of costumes, scripts, and props allows visitors to embody Barrie's literary and dramatic output.

As part of Moat Brae's new extension, the National Centre for Children's Literature and Storytelling promotes literacy and storytelling and provides activities for visitors and educational groups (Moat Brae, 2022). This site therefore offers a multi-sensory experience which extends the existing biographical link with the author and his creation. Meanwhile, 150 miles away lies the small, whitewashed cottage where the author was born in 1860 and spent his early childhood, now cared for by the National Trust for Scotland. Even further lies what is perhaps the most famous literary site connected to Barrie – the Peter Pan statue in Kensington Gardens, London where the magical, fictional boy lived. Surrounded by depictions of squirrels, rabbits, mice, and fairies, and in a bosky situation by Long Water, the bronze statue was commissioned by Barrie himself in 1912. Ever the dramatist, he arranged for it to be erected overnight so that Peter appeared magically to children walking in the park on the morning of the first of May. The statue has been a popular site within the park ever since. The above examples illustrate the process of extending a literary theme through the deliberate creation of visitor attractions based on the life or work of an author. In the following two very famous and oft-cited cases, sociocultural construction reaches its apogee where places that have hitherto only existed within the imaginative realm are fully realised in concrete form, as perfect Baudrillardian *simulacra*.

The fictional environment made real: From Westward Ho! to Baker Street

In an article entitled 'Towns Made by Books,' published in the *English Illustrated Magazine* in 1901, George Wade reviews a number of inhabitations which had become well known because of a literary link. In this essay, he refers to tourists' appreciation of the English towns of Bedford, Kenilworth, and Cromer, indicating that their days of glory had come and that surely in time Haworth might join them in this fame, a clear indication that the longevity of literary repute cannot be easily predicted. However, more salient to this chapter is the opening section of Wade's essay which begins with a description of a town that really was brought into being by a book – the seaside resort of Westward Ho! The name comes from the title of Charles

Kingsley's 1855 novel *Westward Ho!* which was set in nearby Bideford in Devon, in the West of England. The stirring tale, set during the reign of Elizabeth the First, traces the adventures of local lad Amyas Leigh as he follows Francis Drake on a sea voyage. This vivid narrative brought many visitors to Bideford which embraced them 'and began to raise another town for their accommodation, which, in honour of the novel that had caused all this, was named Westward Ho!' (Wade, 1901, p. 63). Wade describes this new town's fine houses and impressive hotels, gardens, and colleges built by the Northam Burrows Hotel and Villa Building Company which was founded in 1863. The Westward Ho! Hotel was reluctantly opened by Charles Kingsley in 1864 but he was said never to have returned to the resort as he believed the development would spoil the beauty of the Devonshire coastline. Apart from being a town whose very existence derives from a work of fiction, the resort also has another famous literary connection. In 1878, Rudyard Kipling joined the new United Services College, a boarding school for children of officers posted overseas, which had recently opened in Westward Ho! The Head teacher recognised his writing skills early on and he was appointed editor of the school magazine (Westward Ho! History Society, 2022). The resort today is still popular although the stately villas now look down upon caravan parks, holiday villages, and amusement arcades. The resort map proudly boasts that the town is the only one in existence with an exclamation mark in its name (see Fig. 6.1).

Figure 6.1 Street map of Westward Ho! in Devon, Southwest England. Photograph by the author.

One of the most famous addresses in English literature, 221b Baker Street, London, home of Consulting Detective, Sherlock Holmes, is another example of a fictional locale made real. We have already seen in Chapter Five that Holmes is in many ways a more fully rounded figure than his creator, Sir Arthur Conan Doyle, and this is augmented by his fictional home which has been created within an existing 1815 Georgian townhouse. When Conan Doyle first began writing his detective tales, the buildings in Baker Street in London did not go up to 221 so the actual address never existed as a domestic dwelling. When the street was extended, the numbers did increase but 221 was subsumed within the address for a branch of the Abbey National Building Society which was to be found at 219–229 Baker Street (Evans, 2019). The Abbey National occupied the site from 1932–2002 and employed a secretary to answer the many letters from Sherlock fans which were sent there. The Sherlock Holmes Museum was set up as a commercial venture in 239 Baker Street and opened its doors to the public in 1990, exhorting fans to 'come and immerse yourself in the gas-lit world of London's iconic detective' (Sherlock Holmes Museum, 2022). The Museum boasts a faux blue plaque and the street number, 221b. This was not, however, the first attempt to recreate the entirely fictional dwelling as a mock-up of Holmes and Watson's sitting room was produced as part of the Festival of Britain in 1951 and subsequently travelled to New York and Toronto before finding a permanent home in the Sherlock Holmes public house in Northumberland Street in London (Booth, 2017). Evans comments on the lack of physical description of the rooms and their furnishings within the texts themselves, although the iconic items of the violin, the Persian slipper stuffed with tobacco, and the dagger spearing unopened correspondence to the chimneypiece are carefully described and faithfully rendered in both the Museum and the Festival of Britain exhibit (Evans, 2019). The Museum has been further concretised by the use of its exterior in the very popular BBC television series *Sherlock* which first aired between 2010–2017. The site therefore attracts both literary and screen tourists along with the dedicated members of the North American *Sherlockian* and British *Holmesian* societies who, for over a century have created a fandom which treats Watson as the true author of the stories, Conan Doyle as the literary agent, and the characters as real people (Pearson, 2007).

The examples of Westward Ho! and 221b Baker Street can, of course be understood purely in commercial terms – their developers saw an opportunity to create a site which they knew would be popular as the intended audience had already been identified through the readership of the associated work. However, the imaginative co-creation of these socially-constructed sites with literary tourists who have already brought to the simulacrum their own knowledge and experience of the fictitious locations means that they can never be seen solely as a base commodification of the genius which inspired them. And for those without prior literary knowledge, there's a wide sandy beach or a well-stocked gift shop to enjoy.

Producing the literary theme

Socially-constructed literary tourism attractions are, in most cases, commercial enterprises which utilise a literary theme to create a unified leisure experience. There are also examples of such attractions which have been created by local authorities in a bid to develop the tourism product in their region. These sites may take the form of replicas of established literary sites reproduced in very different contexts to the original. Such replicas, with their similarity to the constructions found in sites of consumption such as casinos, cruise ships, shopping malls, and theme parks, may attract disdain from those that seek a more authentic communion with literary genius. However, the skill and verisimilitude with which some replicas are produced, combined with the will of the visitor to experience them as the real thing, may produce an existential authenticity that is entirely satisfying to them. Thus, whilst the types of literary location that we have discussed in the preceding chapters attract the literary fan and interested reader, themed sites are aimed at a wider audience who may be in the area for a variety of reasons. Thus, literary locations 'are no longer exclusive places for those seeking enrichment by experiencing the settings of the author's works or working environments. Instead, they have been transformed to tourist destinations catering to the ample needs of the modern tourist' (Müller, 2006, p. 214). These sites use stories and characters from literature to create an attractive narrative that can be enjoyed by all visitors, not just those well-versed in the life and works of particular authors (see Box 6.1). The literary narrative provides a theme which can run through the entire site, including ancillary services, hospitality, and catering.

Box 6.1 Case study: Socially-constructed literary sites in Japan

Replica literary attractions developed in new locations are prime examples of the socially-constructed literary site. These may be replicas of real places or constructions of settings which have only ever existed in the imagination of the author and reader, what Baudrillard termed a *simulacrum*. Both types of attraction evidence the popularity of particular literary figures and themes and, on a practical level, provide access to places that may otherwise be difficult to reach. This is the case in Japan where several constructed sites are found supplementing the Japanese traveller's very evident enthusiasm for the real thing.

In an article on the Japanese affinity with British children's literature, Catherine Butler describes the range of associated literary visitor attractions to be found in Japan. These tend to be constructed using a version of Britishness which focusses on small, quaint hamlets, stone-built cottages, and country manor houses, all of which take their cue from the United Kingdom (UK) region known as the Cotswolds, which is situated between Oxfordshire and Gloucestershire. The reasons for this remain obscure although Butler traces this rapport to Emperor Hirohito who stayed in the Cotswold village of

Bibury during a tour in 1921 and allegedly echoed William Morris in declaring it his favourite place in Britain (Butler, 2020 p. 204). This opinion has been replicated in Japanese guidebooks to England ever since. Aligned to this affinity is the Japanese love of British children's literature such as *Peter Rabbit, Alice in Wonderland*, and the *Harry Potter* series. Consequently, there are several replica Cotswold villages to be found in Japan wherein visitors can enact 'British' behaviour such as taking afternoon tea and being served by butlers, within environments which utilise literary themes. Thus, in attractions such as Yufuin Floral Village in Kyushu, Shuzenji Niji no Sato in Shizuoka Prefecture, or British Hills in Fukushima Prefecture we find shops, restaurants, and activities themed with reference to *Alice in Wonderland* and *Peter Rabbit* along with a clear Hogwarts aesthetic in some buildings.

The Japanese love of Anglophone and European literature is clearly demonstrated in other visitor attractions which replicate literary tourist sites found overseas. The Canadian World attraction in Ashibetsu which opened in 1990 features a replica of the *Anne of Green Gables* House on Prince Edward Island (the original house itself being a construction of a fictional location) along with other buildings such as the school and railway station which were important parts of L.M. Montgomery's fictional seaside town of Avonlea. The commercial attraction closed down in 1997 but was taken over by the local government and is now maintained as a municipal park with the buildings still intact and staffed by volunteers.

Near Tokyo can be found the Maruyama Shakespeare Country Park with its cluster of half-timbered houses, including the Birthplace, Mary Arden's house, a tearoom, Elizabethan garden, a maypole, and a miniature theatre. In the pursuit of authenticity, English oak was imported to build this British architect-designed attraction using traditional methods. Seargeant (2005) comments that this village may claim to be an improvement on Stratford-upon-Avon as it also boasts a reconstruction of New Place, Shakespeare's final home, the original of which no longer exists, having been destroyed by fire in 1759.

In turning works of fiction and the lives of novelist into a leisure theme, these attractions are reflecting a trend that has been discernible since the late 1990s. Lukas (2016, p. 3) suggests that in the last 20–30 years, 'themed space has been the subject of widespread analysis and criticism in academic communities as well as a popular source of entertainment for people around the world. Themed spaces have, in their foundation, an overarching narrative, symbolic complex, or story that drives the overall context of their environs.' Such theming can be found in a variety of different contexts, for example Shaw and Williams (2004) propose a typology of themed spaces: in their classification, *Themed Environments* include shopping malls, restaurants, marketplaces, museums, and galleries, *Theme Parks* incorporate nature, heritage, and cultural attractions, and *Themed Landscapes* describe spaces that have been purposefully created by tourism developers around popular cultural themes such as literature, films, and TV. The pervasive narrative essential to all themed environments indicates that fictional tales, characters, and authors can provide a very rich seam of stories and there are numerous examples to demonstrate this. In these highly controlled spaces 'the visitor experience, interpretation, attractions and hospitality are incorporated' (MacLeod, 2017, p. 425) and the most famous creator of such leisure environments, Walt Disney, is constantly invoked and sometimes used as a critique as we shall see later in the chapter.

Dickens World in Chatham, Kent was a commercial venture based on the life and works of Charles Dickens. The attraction had some biographical connections in that it was situated near to the author's childhood home from 1817 to 1822 and to his final dwelling, Gad's Hill, where he lived from 1860 until his death 10 years later. The site was also close to the town of Rochester which holds an annual Dickens Festival. Dickens World opened in 2007 and closed in 2016, with large financial losses. When the attraction first opened, it combined theme-park elements such as a *Great Expectations* themed boat ride, a haunted house, and a 4D movie at Peggoty's Boathouse. Museum-style interpretation was also included such as an interactive Victorian schoolhouse where children could be tested on their knowledge of the author and his novels. The theatrical aspects of the author's work were acknowledged in the Britannia Theatre where an animatronic Dickens addressed the audience whilst the disembodied voices of his characters intervened. Visitor amenities included a Fagin's Den play area, the Six Jolly Fellowship Porters Pub, and a gift shop (Hughes, 2010; Fleming, 2016).

There was much critical commentary on the opening of Dickens World with many suggesting that a themed attraction was not the best way to celebrate one of the country's best loved authors. However, Hughes, writing in the *Journal of Victorian Culture* in 2010, did not believe that the attraction trivialised its theme – 'this is not Dickens "lite," a soft-focus sketch of the author's most crowd-pleasing creations such as Fagin or Uriah Heep. Rather you will find references to characters whose details might tax the resources of even the most committed Dickens-phile' (Hughes, 2010, p. 389). Additionally, she advises us that the prestigious Dickens Fellowship acted as advisor on the project. Gould and Mitchell (2010) also focus on the literary intentions of Dickens World and its resistance to the more blatant aspects of commodification in the Olde Curiosity Shoppe, the attraction's retail outlet. Souvenirs bearing the name Dickens World, lavender bags, and tea caddies abounded but there was no attempt to plunder the texts to produce, for example, a Cratchit Family Christmas Turkey platter or Mrs Bagnet umbrella. The authors' minds 'boggle' (Gould and Mitchell, 2010, p. 161) at the thought of an Oliver Twist gruel bowl but just such an item is in fact on sale in the gift shop of Charles Dickens' former home in Doughty Street in London, arguably a more authentic biographical site. It would appear that the more official site feels justified in exploiting the literary output (presumably to raise funds for the museum's upkeep) whilst the commercial attraction tended to distance itself from overt commodification.

Another example of a purpose-built literary attraction exploring the life and works of a famous and much-loved author is the World of Astrid Lindgren theme park in Vimmerby, Sweden. This park was created in 1981 to meet the demand from visitors who wanted to see places associated with the author's most famous literary creation, Pippi Longstocking. Lindgren grew up in the town and, as we learned in Chapter Four, her childhood home is open to the public. Three books about her red haired, freckled, super-strong, nine-year-old heroine were published between 1945–48 and they have had an enduring and international appeal since then. Lindgren is the fifth most widely translated children's author (after Enid Blyton, the Grimm brothers, H.C. Andersen, and Mark Twain) with versions of her books being translated into 70 different languages (UNESCO, 2022). In the World of Astrid Lindgren, her international audience find the characters from the books brought to life in their constructed settings, some of which came from the sets of a film and TV series based on the books which aired in 1969. The park also includes shops, a theatre, restaurants and accommodation, a camping ground, and an exhibition on the author's life. Although attempts have been made to develop other related literary attractions in the region (Müller, 2006), the Park claims to be the home of Pippi Longstocking, emphasising that their visitors can experience the stories in their real settings. Emphasising the theatricality of the experience, the website declares that 'here, you and your family become part of the classic stories, and the line between story and reality, and between theatre and play, is always blurred. Here, Astrid Lindgren's stories come to life' (World of Astrid Lindgren, 2022).

These two examples exemplify the integrated approach to creating a highly themed literary environment exploring the life and work of an individual author. We will now continue the discussion with a consideration of similarly heavily-themed attractions which use literary motifs, particularly the fairy tale and aspects of children's literature in general, to produce leisure experiences.

Away with the fairies: The enduring power of fairy tales in tourism

Children's fiction provides a rich mine of material for developing visitor attractions as tales designed for children are often bold, simple, and extreme, all of which qualities lend themselves to the production of engaging, intergenerational attractions which can easily transcend language and cultural borders and allow for the development of merchandise and themed services (Robinson and Andersen, 2002). Within the field of child-oriented literature, the most oft-used literary theme which is utilised within this social construction of literary sites is, of course, the fairy tale. These well-known stories of princesses, deep forests, turreted castles, wolves, elves, wicked stepmothers, and spells, handed down through generations, form the basis for many family-orientated theme parks as can be seen in the case study of Efteling Park in the Netherlands (see Box 6.1).

Box 6.2 Case study: Efteling Park, the Netherlands

Efteling Park is a Dutch theme park situated in Kaatsheuvel in the province of North Brabant in the Netherlands. The park is one of the oldest theme parks in the world having celebrated its 70th birthday in 2022. It predates the original California Disney park by three years. Fairy tales have always formed the core of its theme as Efteling began its life as a sports park which then evolved into a nature park and Fairy Tale Forest. The park now incorporates six roller coasters, four dark rides, boat trips, a steam railway, a golf course, and a theatre but maintains its original attractions and its emphasis on ancient myths, legends, fairy tales, and folklore. The park is a much-loved institution for Dutch families and in the Netherlands it is often referred to as the 'keeper of fairy tales' (Efteling, 2022).

The original fairy tales presented in the Fairy Tale Forest in 1952 were Sleeping Beauty, Snow White, The Frog Prince, The Magic Clock, The Chinese Nightingale, The Naughty Princess, The Six Servants, The Gnome Village, Mother Holle, and Wee Walter Messenger. These classic fairy tales by European authors, Charles Perrault, the Brothers Grimm, and H.C. Andersen were augmented by the subsequent additions of, *inter alia*, The Little Match Girl, The Emperor's New Clothes, and much-loved Italian character Pinocchio who joined the park as late as 2016. There are now 30 fairy tales presented to visitors throughout the park. Storytelling is also used across the other park features, such as the rides which all have a narrative associated with them, for example the Fata Morgana dark ride which is based on the Arabian Nights tales.

The Efteling Hotel opened in 1992 and extends the park's offering with a series of themed suites. These include Cinderella, Sleeping Beauty, and Little Red Riding Hood suites, underwater, circus, and carousel suites and even a room decked out in Coca Cola's red and white livery. A Hans Christian Andersen suite was introduced in 2005 to celebrate the author's 200th birthday. The park also has two holiday villages where guests can rent rustic tree houses, or cottages nestling in the woods or in sand dunes. Evening entertainment and catering has also been introduced to encourage overnight stays and create a resort-style ambience. The natural environment is an important part of Efteling's attraction and comprises a large proportion of the park's 180 acres.

The park attracted over five million visitors in 2019 and in 2020 was the most visited theme park in Europe with more visitors that Disneyland Paris (Statista, 2022). In 2017, 80% of the park's visitors were domestic but the profile is becoming increasingly international with more visitors coming from the UK and France.

The enduring appeal of fairy tales is summed up on the Park's website where the personal and cultural value of these stories is recognised: as they suggest, to understand why Efteling is so appealing, we must

> 'try to understand the magical nature of fairy tales. Why they are written and what they do to people. Stories are the most important connecting force in people's lives. They keep us together. As a household, as a family and perhaps also as a nation.'
>
> (Efteling, 2022)

Fairy tales are recognised across the world and although the Northern European tales of Charles Perrault, the Brothers Grimm, and Andersen dominate, these stories are thought to have migrated from South and East Asia and the Middle East (Hemme, 2005), altering as they moved to new regions. It was noted above that Andersen and the Grimm brothers are amongst the most widely translated authors of children's literature and Table 6.1 shows a range of contemporary fairy tale themed visitor attractions from Europe and beyond which also demonstrates the geographical reach of these fantastical stories.

Table 6.1 Fairy tale themed visitor attractions

Fairy tale themed Literary tourism sites	*Description*
Tivoli Gardens, Copenhagen, Denmark	Magical, historic pleasure gardens and amusement park that inspired H.C. Andersen to write The Nightingale. The Flying Trunk Ride takes visitors on a journey through 32 scenes from the tales of Andersen.
The Fairy Tale Route, Germany	A 600 km driving route from Hanau, near Frankfurt, to the city of Bremen in the northeast of the country. The itinerary passes through places associated with the Brothers Grimm, including towns in which they lived and worked, and places linked to the fairy tales that they collected.
Fairy Park, Victoria, Australia	A fairy tale themed park which opened in 1959 and includes the classic European fairy tales of Cinderella, The Frog Prince, Jack and the Beanstalk, and Goldilocks and the Three Bears among many others. Mythical creatures such as the unicorn, the three-headed Hydra, and the Hippogryph also feature.
Andersen Park, Shanghai, China	The first of a series of H.C. Andersen theme parks in China. Opened in 2017, the park includes an H.C. Andersen Museum, along with seven 'theme lands' with sports facilities and sculptures based on the works of the Danish writer. An agreement has been signed with the Royal Danish Consulate for the further expansion of 30 Andersen Parks across China.

Tales of fairies date back to pre-literate societies and were originally part of the oral tradition. Although linked to folk tales, legends, and beast fables such as those of Aesop, fairy tales are different in that they were not linked to a specific place or era – hence the habitual and deliberately vague opening line, 'once upon a time.' In his famous essay 'On Fairy Stories,' first given as a lecture at the University of St Andrews in 1938, J.R.R. Tolkien attempted to define fairy stories, which is not as easy as one might imagine as so few of these tales actually include any fairies. He proposes that what characterises these magical stories is the world in which they take place – the land of Faërie: 'The definition of a fairy-story … does not, then, depend on any definition or historical account of elf or fairy, but upon the nature of Faërie: the Perilous Realm itself, and the air that blows in that country' (Tolkien, 1964, p. 16). Although all readers would agree that fairy tales always involve magic, enchantment, and extraordinary events, in the land of Faërie, these tales play out as rational stories with expected outcomes. This is what gives fairy tales their reassuring nature – a world is built wherein the unexpected and the wondrous are anticipated and reliably resolved.

The fairy tale shifted from the oral to the literary tradition in the seventeenth century with the efforts of authors and collectors such as Perrault, Madame d'Aulnoy, the Grimm brothers, and Andersen. French author Charles Perrault (1628–1703) produced retellings of traditional tales specifically aimed at children and originally written for his own offspring. His collection entitled *Tales from Times Past, with Morals (*or more colloquially*, Tales of Mother Goose)* was published in 1695. The volume included eight literary retellings including Sleeping Beauty and Little Red Riding Hood. Fairy tales were written and exchanged in aristocratic salons and the work of French author Marie-Catherine d'Aulnoy (1650/1651–1705) who, it is believed, coined the term *fairy tale* ('contes de fées') was popular. Her tales were of an adult nature and Wilhelm and Jacob Grimm, who collected their stories over 100 years later, also resisted the reworking of existing German tales for a young audience but instead tried to maintain the original plot, characterisation, and style of the stories. However, it was only later when they rewrote tales such as Hansel and Gretel and Rapunzel, removing violence and sexual themes, that they gained widespread popularity as tales for children. Andersen, perhaps the most celebrated teller of fairy tales, published over 150 stories from 1835 until just before his death in 1872. These were retellings of traditional tales as well as new stories and included well-loved tales such as The Ugly Duckling, The Snow Queen, The Princess and the Pea, and, perhaps his most famous, The Little Mermaid, a bronze statue of which stands on Copenhagen's seafront in Denmark.

These tales of wondrous beings in enchanted worlds have entertained generations of young readers but they also had a more didactic function as Powell (2014) explains. He discusses the role of socialisation which traditional European fairy tales have played in the lives of children in the nineteenth century, modelling bourgeois behaviour and values and used extensively in nurseries, schools, and libraries in Europe and America. He also points towards the more recent challenges to these tales of magic, found in Oscar Wilde's *The Happy Prince* (1888) and L. Frank Baum's *Wonderful Wizard of Oz* (1901). Fairy tales are still being written and examples of the genre are among some of the most popular children's fiction of the contemporary era as the works of Phillip Pullman and J.K. Rowling demonstrate. Rowling's boy wizard, Harry Potter, has inevitably conjured up numerous visitor attractions based on the novel series and the subsequent films, ranging from platform 9 ¾ at London's Kings Cross railway station to an entire wizarding world in Orlando, Hollywood, and beyond.

Although one of the key features of fairy tales was their spatial and temporal obscurity, they became increasingly associated with the Medieval, albeit a rather hazy and romanticised version of this historic period. For nineteenth century writers and collectors of fairy tales, the era became a focus for nostalgia and their construction of the Medieval is 'primitive, simplistic and often idyllic' (Gutierrez-Dennehy, 2019, p. 72). This was furthered by pre-Raphaelitism and the Arts and Crafts Movement where Medievalism was equated, in the work of Morris, Rossetti, and Byrne-Jones, with the authenticity and honesty of an artisanal, pre-industrial world. The Medieval still has a strong hold over our collective imaginations and it is the favoured historic environment for visitor attractions focussing on fairy tales and indeed for the presentation of historic cities (Lovell, 2019).

Medievalism was also, of course, the motif used by the Disney corporation as a backdrop to their interpretation of European fairy tales, both in their animated films and in the theme parks. Disneyland's Fantasyland, which is one of the park's themed areas, is replicated across the Disney sites of the world and features a Medieval castle, Arthurian attractions, and the specific Disney characters which are based on original fairy tales. Thus, Sleeping Beauty, Snow White, and Cinderella feature heavily along with chivalrous knights, dragons, and damsels in distress. This highly romanticised version of the Medieval period offers historicised escapism, devoid of any reference to the darker aspects of the time such as the Crusades or the Black Death (Gutierrez-Dennehy, 2019). The security and predictability of the theme park concept allows children to safely encounter the magical and the exotic (Karis, 2018) whilst providing nostalgia for parents who make up a large proportion of the Disney audience (Bryman, 1999). The associated merchandising, costumes, experiences, catering, accommodation, and performance further integrate the fairy tale theme in Fantasyland suggesting that 'perhaps the ultimate way of packaging literature is represented by purpose-built attractions: theme parks and heritage parks that have sought to condense elements of imaginative writing … within a created, controlled and largely artificial environment' (Robinson and Andersen, 2002, p. 23). Walt Disney is seen as the forerunner of this process and the term 'Disneyization' has been coined to denote recent changes within the built environment which mimics the Disney process. Thus, places that are themed, where pleasure, profit, and customer service are integrated, where the focus is on the sale of merchandise and where workers perform emotional labour, exhibit the principles of 'Disneyization' (Weaver, 2006). The term is often used in a pejorative manner although the considerable skill with which the Disney corporation continues to expand its empire is acknowledged. Within the realm of literature, 'Disneyisation' refers to Disney's attempts to reframe the classic European fairy tales within an American lens, emphasising both the past and the present and suggesting that anyone can become a princess or a king (Gutierrez-Dennehy, 2019), particularly if they buy the associated merchandise.

Conclusion

In this chapter we have seen that not all sites of literary fiction tourism are determined by the author, neither being immediately linked to their biography nor the settings for their work. Some very popular sites have instead been deliberately constructed in order to employ a literary theme, either *in situ* or as part of a more free-floating homage to an author, a literary work, or a genre. Sites which *extend* the literary theme may provide additional experiences such as visitor centres situated at a biographically authentic literary location. Such centres offer a multi-sensory experience for visitors, involving visitor participation and mediatised experiences such as the Vargas Llosa Museum and Moat House. On the other hand, a socially-constructed site may *produce* the literary theme by offering a decontextualised attraction using a literary motif on which to base a more diffuse experience such as that found in theme parks where fairy tales often form the basis of the park's narrative. These sites are not predicated on authorial biography or geographical accuracy but on the pleasure to be found in stories and in their hyper-real reconstructions.

Socially-constructed literary sites are inevitably more prone to attracting criticism for their commodification and lack of authenticity than other types of literary setting presented to visitors. Literary pilgrims worshipping at the shrine of a favourite author may be scornful of the themed, replicated, and commercially successful sites that have been described in this chapter. However, we have seen that these attractions offer a literary-inspired experience within a controlled setting to a wide range of visitors for whom objective authenticity may come second to enjoying a pleasant day out. Any exposure to literary themes may 'expand the insight of the visitor into the art and craft of writing and that the world of literary art gains rather than loses through the process' (Robinson and Andersen, 2002, p. 15).

The presence of these socially-constructed literary sites underpins the potency of the literary marker as a unifying theme to create place identity. In the following chapter we will explore the role of literary heritage in towns and cities which have used these links to create cultural capital, thus establishing or reaffirming themselves as places of literary repute. Literary festivals, bookshops, and other markers of literature and literacy in towns and cities will also be explored.

References

Baudrillard, J. (1983) *Simulacra and simulation*. New York: Columbia University.

Booth, A. (2017) 'The detective annex: Where is 221B Baker Street?', *Victorian Review*, 43 (1), pp. 1–6. Available at: https://www.jstor.org/stable/26809595 (Accessed: 1 November 2022).

Brooks, L. (2019) 'J.M. Barrie's "Peter Pan house" to open as children's literature hub', *The Guardian*, 27 May. Available at: https://www.theguardian.com/books/2019/may/27/jm-barries-peter-pan-house-to-open-as-childrens-storytelling-hub-in-scotland (Accessed: 11 October 2022).

Bryman, A. (1999) 'The Disneyization of society', *The Sociological Review*, 47 (1), pp. 25–47. Available at: https://journals.sagepub.com/doi/epdf/10.1111/1467-954X.00161 (Accessed: 2 June 2023).

Butler, C. (2020) 'Japan reads the Cotswolds: Children's literature, tourism, and the Japanese imagination', *Children's Literature*, 48 (1), pp. 198–233. Available at: https://doi.org/10.1353/chl.2020.0008 (Accessed: 15 October 2022).

Efteling (2022) Efteling Theme Park Resort. Available at: https://www.efteling.com/en?gad=1&gclid=EAIaIQobChMIpoOZstmagAMV0n9MCh00_geIEAAYASAAEgK4yPD_BwE (Accessed: 3 September 2022).

Evans, M.K. (2019) '"The Name is Sherlock Holmes, and the address Is 221B Baker Street": Virtual reality, fan communities, and tourism', *The Journal of Popular Culture*, 52 (6), pp. 1494–1511. Available at: https://doi.org/10.1111/jpcu.12870 (Accessed: 4 November 2022).

Fleming, P. (2016) 'After Dickens World: Performing Victorians at the Chatham Docks', *Neo-Victorian Studies*, 9 (1), pp. 13–31. Available at: https://neovictorianstudies.com/issue/view/20 (Accessed: 10 November 2022).

Fox, R. (2008) 'Marin Držić: A case for Croatian literary tourism', *Journal of Heritage Tourism*, 3 (2), pp. 138–153. Available at: https://doi.org/10.1080/17438730802138154 (Accessed: 12 October 2022).

Gould, M. and Mitchell, R.N. (2010) 'Understanding the literary theme park: Dickens World as adaptation', *Neo-Victorian Studies*, 3 (2), pp. 145–171. Available at: https://neovictorianstudies.com/article/view/242 (Accessed: 18 September 2022).

Gutierrez-Dennehy, C. (2019) 'Taming the fairy tale: Performing affective Medievalism in fantasyland', in Kokai, J.A. and Robson, T. (eds.) *Performance and the Disney theme park experience*. Cham: Palgrave Macmillan, pp. 65–83.

Hemme, D. (2005) 'Landscape, fairies and identity: Experience on the backstage of the fairy tale route', *Journal of Tourism and Cultural Change*, 3 (2), pp. 71–87. Available at: https://doi.org/10.1080/09669580508668488 (Accessed: 28 September 2022).

Herbert, D. (2001) 'Literary places, tourism and the heritage experience', *Annals of Tourism Research*, 28 (2), pp. 312–333. Available at: https://doi.org/10.1016/S0160-7383(00)00048–00047 (Accessed: 2 June 2022).

Hughes, K. (2010) 'Dickens world and Dickens' world', *Journal of Victorian Culture*, 15 (3), pp. 388–393. Available at: https://doi.org/10.1080/13555502.2010.519531 (Accessed: 14 May 2022).

Karis, M. (2018) 'Between the lines: children's literature and the Disney theme parks', *Journal of Themed Experience and Attractions Studies*, 1 (1), pp. 69–71. Available at: https://stars.library.ucf.edu/jteas/vol1/iss1/9 (Accessed: 13 August 2022).

Lipovšek, E. and Kesić, S. (2015) 'Commodification of culture in fiction-induced tourism', *Tims. Acta: naučni časopis za sport, turizam i velnes*, 9 (2), pp. 105–113. Available at: https://doi.org/10.5937/timsact9-8125 (Accessed: 28 October 2022).

Lonely Planet (2022) Lonely Planet Guide Peru: Casa Museo Mario Vargas Llosa. Available at: https://www.lonelyplanet.com/peru/arequipa-and-canyon-country/arequipa/attractions/casa-museo-mario-vargas-llosa/a/poi-sig/1604234/363389 (Accessed: 13 December 2022).

Lovell, J. (2019) 'Fairytale authenticity: historic city tourism, Harry Potter, medievalism and the magical gaze', *Journal of Heritage Tourism*, 14 (5–6), pp. 448–465. Available at: https://doi.org/10.1080/1743873X.2019.1588282 (Accessed: 20 December 2022).

Lukas, S.A. (2016) 'Introduction: The meanings of themed and immersive spaces' in Lukas, S.A. (ed.) *A reader in themed and immersive spaces.* Pittsburgh, PA: ETC Press, pp. 3–15.

MacLeod, N. (2017) 'The role of trails in the creation of tourist space', *Journal of Heritage Tourism*, 12 (5), pp. 423–430. Available at: https://doi.org/10.1080/1743873X.2016.1242590 (Accessed: 4 March 2021).

Moat Brae (2022) Moat Brae National Centre for Children's Literature. Available at: https://www.moatbrae.org/ (Accessed: 10 December 2022).

Müller, D.K. (2006) 'Unplanned development of literary tourism in two municipalities in rural Sweden', *Scandinavian Journal of Hospitality and Tourism*, 6 (3), pp. 214–228. Available at: https://doi.org/10.1080/15022250600667433 (Accessed: 11 October 2022).

Pearson, R. (2007) 'Bachies, Bardies, Trekkies, and Sherlockians', in Gray, J., Sandvoss, C. and Harrington, C. (eds.) *Fandom: Identities and communities in a mediated world.* New York: NYU Press, pp. 98–109.

Powell, F. (2014) 'The cultural landscape of childhood: Fairy tales, little people and media giants', *An Leanbh Óg*, 8, pp. 23–38. Available at: https://omepireland.ie/wp-content/uploads/2019/07/An-Leanbh-Og-Volume-8-2014.pdf#page=31 (Accessed: 4 January 2023).

Robinson, M. and Andersen, H.C. (2002) 'Reading between the lines: Literature and the creation of touristic spaces' in Robinson, M. and Andersen, H.C. (eds.) *Literature and tourism: Essays in the reading and writing of tourism.* London: Thomson, pp. 1–38.

Seargeant, P. (2005) 'Globalisation and reconfigured English in Japan', *World Englishes*, 24 (3), pp. 309–319. Available at: https://doi.org/10.1111/j.0083-2919.2005.00412.x (Accessed: 18 December 2022).

Shaw, G. and Williams, A.M. (2004) *Tourism and tourism spaces.* London: Sage.

Sherlock Holmes Museum (2022) About the museum. Available at: https://www.sherlock-holmes.co.uk/about-the-museum/ (Accessed: 4 June 2022).

Statista (2022) Attendance at amusement and theme parks in Europe by facility from 2019 to 2021. Available at: www.statista.com/statistics/194291/attendance-at-theme-and-amusement-parks-in-europe/ (Accessed: 3 October 2022).

Stiebel, L. (2004) 'Hitting the hot spots: Literary tourism as a research field with particular reference to KwaZulu-Natal, South Africa', *Critical Arts*, 18 (2), pp. 31–44. Available at: https://doi.org/10.1080/02560240485310151 (Accessed: 11 August 2022).

Tolkien, J.R.R. (1964) 'On fairy-stories' in *Tree and leaf.* London: Unwin, pp. 11–70.

UNESCO (2022) Index translationum. Available at: https://www.unesco.org/xtrans/bsstatexp.aspx?crit1L=5&nTyp=min&topN=50 (Accessed: 28 October 2022).

Wade, G.A. (1901) 'Towns made by books', *The English Illustrated Magazine*, Issue 211, pp. 63–70.

Watson, C. (2013) 'Tourist trap: Re-branding history and the commodification of the South in literary tourism in Mississippi', *Studies in Popular Culture*, 36 (1), pp. 145–161. Available at: https://www.jstor.org/stable/23610156 (Accessed: 27 July 2022).

Weaver, A. (2006) 'The Disneyization of cruise travel', in Dowling, R.K. (ed.) *Cruise ship tourism*. Wallingford: CABI, pp. 389–396.

Westward Ho! History Society (2022) Westward Ho! history. Available at: http://www.westwardhohistory.co.uk/ (Accessed: 3 June 2023).

World of Astrid Lindgren (2022) Welcome in to the stories. Available at: https://astridlindgrensvarld.se/en/ (Accessed: 10 June 2023).

7 Places of literary reputation

Thus far in this book, we have explored a range of literary fiction sites associated with different aspects of a novelist's life or output. These sites are largely individual attractions which have been preserved or developed for a variety of purposes, whether that be commemorative, educational, or commercial. In this chapter we will take a wider view of the tourism potential of authors and their works by exploring places of literary repute. These types of literary settings attract visitors because of their reputation as places that have welcomed and nurtured writers, and which are well-known for producing a strong literary output. These are locations which are considered to have a high level of literary cultural capital (see Chapter Three) which might come from their reputation for hosting literary events or festivals that may celebrate local authors or the area's connection with wider literary themes. They may be renowned for the number of independent bookshops to be found there or for their libraries and literary museums. They may even be towns and cities with well-established publishing houses and associated industries such as bookbinding. All this literary activity may make itself present within the urban landscape through the presence of commemorative plaques, statues and other literary monuments, literary pubs, cafés, and hotels and activities such as themed trails and guided tours. Thus, cities are moving away from 'acting as individualized and one-stand literary attractions based on homage to a writer's location or setting of a work of fiction and becoming more broad and varied in their literary tourism expressions' (Marques, 2019, p. 67). Towns and cities can of course combine all the features described above and, in these cases, they may become a member of the 'book town' network or even be eligible for the United Nations Educational, Scientific and Cultural Organisation (UNESCO) sponsored accolade of City of Literature, the ultimate marker of literary reputation. In this chapter we will examine these settings with a literary reputation and consider the role that this link to the world of books, authors, and readers plays in creating a sense of place identity and in enhancing the cultural capital of both the location and participants.

DOI: 10.4324/9781003041740-7

Literary cities

The concept of the literary city acknowledges the important role that literature plays within the cultural and economic dynamics of urban spaces (Rodrigues de Sousa, 2022) and the literary can manifest itself in many different ways within a city setting. As we saw in Chapter Six, the city plays an important role in the development of literary locations and in the experience of fictional landscapes. We may feel that, as first-time visitors to certain cities, we know them already, through our fictional encounters which have taken us down their streets and alleyways in our minds. Similarly, wholly invented fictional conurbations can be experienced in the flesh when we encounter similar settings in our real-world travels. The real city concretises our imaginings of fictional urban spaces.

Cities also play an important role as the home environments of authors and their inspiration along with hosting the book shops, libraries, publishing houses, and literary events which support the production and consumption of fiction. The literary DNA of cities can be evidenced in the plaques and markers which are fixed onto buildings to denote previous famous residents (for example see Box 1.1 for an overview of London's Blue Plaque scheme) and in street names celebrating authors and characters. In London, one can encounter Crusoe Mews in Stoke Newington, named after one-time resident Daniel Defoe's famous creation and, inevitably, there is a Sherlock Mews leading off Baker Street. References to Charles Dickens' characters (e.g., Little Dorrit Court) can be found in the street names of Southwark, South London which provided the setting for many of his novels. Indeed, according to the Royal Mail in a recent study, 'Copperfield' is the most common novel-based street name in the United Kingdom (UK), featuring in 23 different postcode areas (Royal Mail, 2020) and demonstrating the level to which such much-loved characters are embedded in the cultural and physical landscape. The idea that such literary street-naming may be used to convey cultural capital was satirised in the British Broadcasting Corporation (BBC) situation comedy *The Fall and Rise of Reginald Perrin* (1976–1979) where the eponymous hero's banal daily commute took him along Wordsworth Drive, Tennyson Avenue, and Coleridge Close in the fictional South London suburb of Climthorpe. Perrin's rather drab home on the 'Poets' Estate' was far removed from the sublime environments created by those particular poets.

The quality and prevalence of bookshops is another marker of a literary city, and their existence may in fact promote tourist trips in their own right. Certainly, world famous bookshops such as Shakespeare and Co. in Paris and San Francisco's City Lights have long been on the itineraries of city visitors. Rodrigues de Sousa (2022) notes that 'bookshop or bookstore tourism refers to a type of literary tourism focused on visiting unique independent bookstores and experiencing a travel destination as a book lover' (p. 13). Robertson and Yeoman (2014) note that 'consumers from advanced consumer societies talk about three things more than most, namely books, holidays and food' (p. 321) and the recent appearance of visitor guides to book shops (e.g., Strauss' publication *Do You Read Me? Bookstores Around the World* or Boland's guide to *Bookshop Tours of Britain,* both published in 2020) suggest that this is indeed a popular activity for city break visitors. A city's ability to support a healthy supply of bookshops is another indicator of literary prestige and a means of promoting its intellectual and creative life to the wider world. This chapter will continue to explore places of literary reputation by further examining literary markers within the urban landscape – thus, monuments, museums, parks and literary pubs, cafés, and hotels will be visited with a view to examining their contribution to a literary sense of place.

Literary monuments

Statues and monuments commemorating great leaders, historical events, and illustrious individuals have long been a feature of the urban landscape. They represent the pride of a nation or city in those that once resided there, and literary figures are represented just as frequently as great military or religious leaders. Smith notes that 'monuments have always been useful promotional tools for cities, employed both in traditional advertising literature and as a more subtle form of place marketing' (Smith, 2007, p. 79). We have already considered the creation of Poets' Corner in Westminster Abbey where, since the 1500s, the literary canon is represented in stone and marble for the edification of visitors, shifting the consumption of literature from a private, readerly activity to allow a 'relatively broad range of social classes to congregate together, as a notionally singular public, before the monuments of national literary fame' (Connell, 2005, p. 560). Indeed, in the eighteenth century, Poets' Corner was the subject of tours and guidebooks, and the mass production of replica literary monuments in plaster, lead, or porcelain, depending on budget, is evidence of the significance of the author within the culture of the city and the nation (Connell, 2005).

Statues and monuments are placed in the public sphere to encourage commemoration – a call to remembrance which, in its appeal to collective memory, is much more socially and culturally situated than the simple recording of historical fact. Thus, statues are consumed within the context of the viewers' own consumption of media, tourist sites, and their own personal histories and traditions (Savage, 2007). Larger than life and set up high on plinths in places where they can be easily viewed, statues and other commemorative monuments were originally conceived to convey an aura of greatness from a distance and to promote a monumental gaze. The statue of Malta's national poet Dun Karm Psaila (1871–1961) which stands in a prominent position at the entry to the city of Valletta is a good example of this traditional approach (see Fig. 7.1). The life-size bronze statue is positioned on a 4-foot-high marble base placed on a set of five steps and his literary credentials are depicted by the sheaf of papers that he holds in his hand.

Figure 7.1 Statue of Dun Karm Psaila in Valletta, Malta. Photograph by the author.

Perhaps the most impressive mementos of literary greats are the world's two tallest monuments to writers. The highest is the José Martí Memorial in the centre of Plaza de la Revolución Square in Havana, Cuba. The memorial was unveiled in 1958 and incorporates a star-shaped tower, 109 metres high and a statue of the writer surrounded by columns. The second tallest is the gothic spire of the Scott Monument in Edinburgh which stands at 61 metres high on the city's main thoroughfare overlooking Edinburgh Castle. The spire contains a seated marble statue of Sir Walter Scott and his dog Maida, and 68 smaller figures of his fictional characters adorn the monument which also houses a museum. Next to the Scott Monument can be found Waverley Station, the world's only railway terminus named after a novel, having been given its name in 1854 in recognition of the significant amounts of train travel that Scott's novels promoted. Both the Scott Monument and the José Martí Memorial are spectacular edifices that proclaim the pride and centrality of the authors and their legacy to their cities and are visitor attractions in their own right.

Despite the importance of scale and monumentality in depicting revered authors within the city, literary statues can also be of a modest nature and encourage a more personal engagement such as the statue of Hans Christian Andersen in Central Park, New York. Here, the monumental gaze is replaced by a more embodied experience of the literary figure. Andersen is depicted in bronze, seated on a granite bench, and reading from his book *The Ugly Duckling*, with an attentive duck as his audience. The sculpture was designed to be handled and every Saturday throughout the summer since it was first erected by the Danish American Women's Association in 1956, storytellers read Andersen's stories to children who gather by the foot of the statue. A nearby statue of Alice in Wonderland, accompanied by the Mad Hatter, White Rabbit, and Cheshire cat, was also designed for children's interaction and parts of the bronze sculpture have been rubbed smooth over the years by their feet and hands. Bringing statues to life is the motivation behind the National Lottery-funded *Talking Statues* project from the events company, Sing London. Through this initiative, 30 statues in London can converse with their viewers at the swipe of a Quick-Response (QR) code. Voiced by well-known actors, the statues will talk about their life in the city by means of a call delivered directly to the user's mobile phone which is intended to create a personal communion between the statue, the life it represents, and the viewer. Five of the statues are of literary figures, namely John Keats, William Shakespeare, George Orwell, Peter Pan, and Sherlock Holmes.

Literary museums and libraries

In Chapter Four, we explored in detail the evolution of the writer's house as a means of portraying the life and work of the author. As Young (2015) explains, this is the most common way in which the life of a literary figure becomes museumised as the intimate domestic setting provides both biographical detail and tells us much about the individual author's inspiration. Museums, on the other hand, were originally developed to be public spaces devoted to upholding national culture and promoting wider education, although the focus has now shifted to the representation of more devolved identities. Despite the prevalence of individual writers' houses, there are also examples of specific literary museums which have been created to allow cities to explore their local and national literary heritage. The Writers' Museum in Edinburgh which celebrates Roberts Burns, Robert Louis Stevenson, and Sir Walter Scott (all of whom have houses dedicated to them elsewhere) is run by Edinburgh City Council. The museum stands beside Makar's Court, a public square dedicated to memorialising the words of Scotland's literary greats, carved into stones. According to Edinburgh City council, 'Makars' Court is an evolving national literary monument, with new stones added regularly' (Edinburgh Museums, 2023) which suggests that the creation of national foci for public literary appreciation is still a current preoccupation. This is borne out by the opening of the Museum of Literature, Ireland in Dublin in 2019. This institution is a partnership between University College Dublin and the National Library of Ireland with funding and guidance from Fáilte Ireland, the national tourist authority (MoLI, 2023). MoLI's mission is to interpret Irish storytelling from the earliest traditions to the work of contemporary writers, thus firmly placing literature at the heart of national Irish heritage. Similarly, the American Writers' Museum in Chicago, which is also a recent creation having opened its doors to the public in 2017, aims 'to engage the public in celebrating American writers and exploring their influence on our history, our identity, our culture, and our daily lives' (American Writers' Museum, 2023).

Museums are institutions that represent national and local culture via public education, and they are joined in this mission by the library. Amongst the oldest institutions in the world, libraries have long been associated with tourism, with early Grand Tour itineraries featuring them as both visitor attractions and as places to learn about the destination, borrow books for leisure reading, and meet like-minded travellers (Tokić and Tokić, 2018). Today, libraries continue to be of interest to tourists, with some researchers referring to this relationship as 'library tourism' (Lainsbury, 2019) or 'bibliotourism' (Roque and Guerreiro, 2021). With many being housed in impressive, ornate, and ancient buildings, the most famous libraries feature in published lists of most-visited attractions. The Association of Leading Visitor Attractions reported that the British Library in London and the Bodleian Libraries in Oxford were respectively the 20th and 37th most visited locations in the UK in 2022 (ALVA, 2023). Such well-visited, iconic buildings may be ancient as in the case of the Vatican Library (founded in 1475) or modern as in the contemporary version of the Alexandria library in Egypt, opened in 2002 and based on the original classical wonder of the ancient world (see Table 7.1).

Table 7.1 Famous libraries as visitor attractions

Library	*Visitor facilities available*
The Library of Congress, Washington, DC	Founded in 1800, the Library of Congress is the largest library in the world housing more than 32 million books, 61 million manuscripts and many historical documents. Visitors can follow a digital guide through the historic buildings.
Reading Room of the British Museum, London, UK	Located in the Great Court of the British Museum, the restored Reading Room is an information centre and houses a collection of 25,000 books, focused on the world cultures represented in the museum. The impressive domed room completed in 1857 was inspired by the Pantheon in Rome.
Bodleian Old Library, Oxford, UK	The Bodleian Old Library dates to 1602 and is part of a group of libraries that supports the University of Oxford. It has three historic reading rooms: the Lower Reading Room, the Upper Reading Room, and Duke Humfrey's Library. Library tours are available to visitors.
Library of Alexandria, Alexandria, Egypt	The original Library of Alexandria was one of the Seven Wonders of the World. This modern version opened in 2002 and the library complex includes eight million books, a conference centre, museums, galleries, a planetarium, and a manuscript restoration laboratory.
State Library of Victoria, Melbourne, Victoria, Australia	This library opened in 1854 as the Melbourne Public Library and is Australia's oldest public library and one of the world's first free public libraries. The founders wished to create a place of learning for all and there is an expanding collection of historical artefacts along with the 4 million items in the library collections.

Whilst architecturally interesting buildings undeniably attract the tourist gaze, the collections themselves can also be a specific draw for the literary and more generalist visitor. In particular, iconic texts and artefacts such as the *Book of Kells* in the library at Trinity College, Dublin, Jane Austen's desk and Shakespeare's first folio in the British Library in London, or the Gutenberg Bible in the Library of Congress, Washington, DC, all attract tourists. The library's spaces may also attract visitors by hosting exhibitions on local history, talks and lectures, films, and theatrical performances. Cultural and literary tourists may be drawn to such places where they can both develop their cultural capital, and pick up reliable, authoritative information about the destination being visited. For many, spending time in a library feels like a civilised thing to do as these are democratic, free spaces which exist for the benefit and progress of society. Lainsbury (2019) refers to them as 'safe spaces' where visitors can escape the noise and confusion of an unknown city and find quiet solace in books and a literary ambience. Of course, churches and cathedrals offer a similar sense of sanctuary, but the added benefit of heating, comfortable seating, cafés, and reliable bathroom facilities add to the library's appeal. It is noted that destination marketing and management organisations have overlooked the touristic draw of libraries and rarely promote them as part of their tourism product (Tokić and Tokić, 2018). However, the hospitality industry has been more active in reflecting the power of the library in its use as a hotel theme as we shall see in the following section.

Literary hospitality: Pubs, cafés, and hotels

In Chapter Four, we noted that public spaces such as pubs and hotels which were frequented by authors can be seen as an extension to the writer's home and have been the focus of literary tourism pilgrimages. Writing is a solitary activity, and we imagine authors taking a break in their local café to people-watch and seek inspiration, or perhaps to meet with like-minded literary companions. For the literary tourist, these spaces may then become as hallowed as the writer's study. Towns and cities with a literary reputation will often promote cafés, restaurants, and hotels which have attracted, sheltered, and nourished authors, demonstrating the crucial role that the city played in the life and output of the author. Many of these hostelries are linked to several different artists, for example *Les Deux Magots* in Paris boasted Simone de Beauvoir, Jean Paul Sartre, James Joyce, and Rimbaud as customers; Milne's Bar in Edinburgh was sanctuary to a group of post-war poets who included Hugh MacDiarmid, Sorley MacLean, and Norman MacCaig; and the White Horse Tavern in New York hosted Norman Mailer, Jack Kerouac, Anaïs Nin, and Dylan Thomas. These literary cafés make much of their heritage and feature in city walking tours and literary pub crawls. A few are even happy to continue to reserve a table for their former literary clientele – Café Tortoni, Buenos Aires has seated wax figures of Jorge Luis Borges and Alfonsina Storni, the Literatunoe Kafe in St Petersburg has a similar figure of

Aleksander Pushkin, and El Floridita in Havana boasts a life-size bronze statue of Ernest Hemingway, propping up the bar.

Literary hotels offer travellers an opportunity to commune with a favourite author, text, or theme for a more extended visit. Certainly, authors use hotels as peaceful retreats or temporary homes and are often inspired to employ them as plot devices as the hotel setting allows for a naturalistic interplay of a wide range of characters, for example in Anita Brookner's *Hotel du Lac* (1984), Agatha Christie's *Evil Under the Sun* (1941), John Irving's *Hotel New Hampshire* (1981), and, most recently, *A Gentleman in Moscow* (2016) by Amor Towles. The Hotel Metropol, in which much of the action of Towles' novel takes place, now offers tours of the rooms and public areas which are featured. Literary themes are increasingly popular motifs for hotels and Çevik (2022) argues that they should be considered literary tourism products in their own right. He distinguishes five separate categories of literary hotel: these include hotels associated with an author's life or dedicated to a favourite writer; hotels with an association with a fictional character or novel; hotels using a more general literary theme; and finally library hotels where the ambience of a bookish environment is created for the benefit of visitors. Many hotels keep a few bookshelves of discarded novels for their guests' use (and it is one of the joys of a holiday to chance upon something interesting therein) but a library hotel will own a large collection of books, rare collections, and even, in the case of the Book and Bed Hotel in Tokyo, offer guests the chance to fall asleep on bookshelf beds, surrounded by books (Çevik, 2022). Certainly, a range of literary-themed hotels will be another feature of a town or city which prides itself on its literary reputation.

Literary placemaking and cultural capital

In the discussion above, we have reviewed a range of features and attractions which are commonly found in places with a literary reputation. These literary cities use the cultural capital that such associations offer to create a sense of place which reflects this heritage and renown. Museums, galleries, and the performing arts will all bestow prestige, but the written word provides an additional stature. We have already seen in Chapter Three that individuals can distinguish themselves through the accumulation of cultural capital, displaying their tastes through institutionalised, embodied, and objective forms of capital. Institutionalised capital is the power we convey through our education and employment status, embodied capital is displayed through our skills, habits, and mannerisms, and objective capital can be identified through our tastes and preferences in material goods including cultural artefacts such as books and leisure activities.

Places that are rich in resources bestowing cultural capital can heighten their status and distinguish them from competitors through marketing these assets (Kostanski and Puzey, 2014). They can also offer experiences which will enable visitors to increase their own cultural standing. It is therefore unsurprising that literary themes are frequently used in city and town placemaking strategies where the literary place image is developed to attract literary tourists and create a cultured identity for residents (Marques, 2019). We have already seen this happen in the creation of literary 'countries' such as the spatial delineation of the lives of W.B. Yeats and Charles Dickens (see Chapter Two). Similar strategies include the production of urban 'literary quarters.' The district in Madrid most associated with writers Cervantes, Quevedo, and Lope de Vega, and a complex of literary sites linked to local authors Dmitry Mamin-Sibiryak and Pavel Bazhov in Yekaterinburg, Russia, have both been given the accolade of 'quarter,' suggested that these are districts where literature dwells. These literary mappings are used by regional and national tourism agencies to promote the touristic potential of destinations and reflect the status to be found in literary cultural capital. However, when the fame of a writer declines, these agencies are not slow to remove the associations as happened in the case of Catherine Cookson Country in the Northeast of England which was created with pride in 1985 but dropped by South Tyneside Council 25 years later. These actions to employ literary themes in placemaking strategies are often taken by tourism agencies and local authorities but they are by no means always a 'top down' process. Robinson and Andersen (2002) note that, instead, 'literary association is a far more organic performance, acted out with various stakeholders proud and astute enough to be associated with their factual and fictional progenies as part of identity building' (p. 21). Thus, local communities, literary societies, and businesses may also make claims for literary associations as indeed will visitors (Ridanpää, 2011). Place is therefore socially constructed by the values it accrues from cultural associations and by the performance of visitors within the setting (Seaman, 2019).

UNESCO Cities of Literature

In this chapter, we have reviewed the numerous literary features, associations, and attractions that coalesce to create a place of literary reputation. Some of these associations have evolved organically and some have been the focus of strategic development by a range of actors. However, the most deliberate harnessing of the cultural capital of a literary place is the UNESCO Cities of Literature project which began in 2004 with the launch of the Creative Cities Network. This network comprises seven creative fields (crafts and folk art; design; film; gastronomy; literature; music; and media arts) and designations are given to cities which 'recognise past, present and future: a strong cultural heritage, a vibrant and diverse contemporary cultural scene, and aspirations to extend culture to the next generation … in a global partnership' (Cities of Literature, 2023). There are currently 42 Cities of Literature in 28 countries, representing 6 continents. Cities bid for the designation and must address a number of criteria. Most Cities of Literature will have a strong publishing and translating industry, educational offerings featuring domestic and foreign literature, a strong presence for literature, drama and poetry, literary festivals and events, plentiful libraries and bookshops, and an active media.

Edinburgh was the first City of Literature in 2004, setting a very high standard, and recently cities such as Vilnius, Beirut, and Bagdad have joined the designated list. The contribution that such titles make to city pride and cultural capital is examined by Kostanski and Puzey (2014) who suggest that such 'toponymic identity' has been used by cities such as Melbourne to attach their intellectual reputation to a globally recognised brand. Cities bidding for the title need to identify the ways in which their various literary resources and agencies can co-operate to increase the potential impact for visitors and host communities (see Box 7.1). Marques (2019) notes that such initiatives represent the way that literature is being used as a strategy for development that goes beyond city branding as 'these efforts correspond to an attempt to create meaning and a storyline which can be used for positioning as a tourism destination but also as a platform where literary resources are foundational to develop creativity and innovation' (p. 68).

Box 7.1 Case study: Durban, South Africa – UNESCO City of Literature

In 2017, Durban became the first African city to be awarded the UNESCO City of Literature status. Durban is the largest city of the KwaZulu-Natal province in South Africa, with over 3,442,400 inhabitants. The accolade recognises the city's proud history of literacy despite the years of apartheid. For example, Durban is home to the prestigious University of KwaZula-Natal (UKZN) which is one of the few universities to offer a PhD in creative writing.

The city's bid for City of Literature was founded on the city's existing literary offerings which included three very successful literary festivals:

- Poetry Africa, an international poetry festival, hosted by the Centre for Creative Arts (CCA) at University of KwaZulu-Natal. The festival programme includes theatre performances, readings, music, and book-launches. The theme for 2022 was Poetic (In)Justice.
- Time of the Writer Festival also hosted by the CCA and now in its 26th year. The 2023 theme is Placemaking: Roots, Influence, Imagination, and Expression
- ARTiculate Africa is South Africa's foremost Literature and Art Fair, hosted by the eThekwini Municipality's Parks, Recreation and Culture unit. This three-day free event includes seminars, exhibitions, and live entertainment and was launched in 2017.

Durban also has an impressive library service with over 100 libraries which run programmes and events throughout the year, including the One City, One Book initiative where a book is chosen by the public and read by all citizens. There is a Reference Library including Rare Books in the Don Africana collection and the Hillier Shakespeare collection. The wider community are served by mobile and branch libraries.

Durban is also home to the Mazizi Kunene Foundation which was founded in 2006. Professor Mazisi Kunene (1930–2006), Africa's first Poet Laureate, was a writer and freedom fighter who was inspired by the history of the Zulu people and wrote his works in the Zulu language. The foundation aims to establish a cultural and literary archive of African writing and primarily to collect, translate, and publish the entirety of Professor Kunene's works (Cities of Literature, 2023).

The KZN Literary Tourism was first set up in 2002 by Professor Lindy Stiebel, of the English Studies department at the University of KwaZulu-Natal. This project explores the links between literature and place and has created an archive of materials related to over 100 writers who have an association with the KwaZulu-Natal province. This archive includes scholarly research, book reviews, local literary events, and interviews with authors and Professor Stiebel is an active researcher in the field of literary tourism. In 2005, KZN Literary Tourism produced a set of eight trails exploring literary landscapes and notable local authors such as Rider Haggard and Alan Paton (KZN Literary Tourism, 2023).

The Durban bid also included plans for the future nurturing of literature in the city. This included a translation programme, collaborations with other cultural providers and events (for example the Durban International Film Festival), and the establishment of statues and trails in the city to further honour its literary heritage.

Book towns

On a smaller geographical scale, but still utilising the power of literary association, is the 'book town', an idea which began in the small historic market town of Hay-on-Wye in Wales, UK. We have already considered the lure of bookshops and the possibility of bookshop tourism, but the book town concept is one based on the deliberate creation of a critical mass of book shops in a single setting in order to regenerate the local economy. This may entail converting properties in declining, peripheral areas into second hand or antiquarian book shops, which then allows the town to develop a recognisable identity which can be packaged for tourism development (Seaton,1999). Hay-on-Wye is set within a picturesque rural setting but in the 1960s had little else to commend it. Entrepreneur Richard Booth set up a large second-hand bookshop in a defunct cinema and with its success, bought up more disused properties until the town became known as a 'book town' in the 1970s. Tourism developed in tandem with an expansion in local hotels, cafés, restaurants, and tea rooms. Associated businesses such as print and map shops, antiques and gift stores also started to appear. New book towns based on the Hay-on-Wye model began appearing across Europe in the 1980s, starting in Belgium, with the small village of Redu (1984), then Bécherel (1988) and Montolieu (1989) in France, followed by Bredevoort

(1993) in the Netherlands (Driscoll, 2018). Scotland's National Book Town, Wigtown, was designated in 1998. A strongly European trend, there are nonetheless book towns in Australia, New Zealand, and Malaysia. This network is designated and supported by the International Organisation of Book Towns network which was set up in 1998 and which defines a book town as 'a small rural town or village in which second-hand and antiquarian bookshops are concentrated. Most Book Towns have developed in villages of historic interest or of scenic beauty' (IOBT, 2023).

Although Mintel (2011) includes book towns as a category in their report on literary tourism, these deliberately constructed retail areas may better resemble theme towns, such as those which become associated with activities such as art or gastronomy (Macleod, 2009). They certainly differ from the book selling districts that have evolved organically, such as the *bouquinistes* along the banks of the river Seine in Paris (now a UNESCO World Heritage site), or those that have historically grown up around educational institutions, such as the Kanda-Jinbōchō district of Tokyo, Japan, and College Street in Kolkata (see Box 7.2), all of which are much-beloved by local bibliophiles and literary tourists alike. However, the book town experience of browsing and buying books in an attractive environment with associated catering, hospitality and retail opportunities is a popular leisure activity for domestic and international visitors who additionally gain cultural distinction from being seen as booklovers. Book towns may therefore offer a 'more diffuse form of cultural tourism' than mainstream literary tourism destinations with their focus on specific authors, characters, or locations (Driscoll, 2018, p. 405).

Box 7.2 Case study: College Street, Kolkata, India

As the world's second largest second-hand book market, College Street is situated in central Kolkata, in West Bengal, India. The street extends along one and a half kilometres and is a gathering place for students, locals, and visitors interested in browsing its many makeshift stalls and bookshops. The street is named after the original Hindu College that was set up there in 1817 but the street is also home to the University of Calcutta, Presidency University, Calcutta Medical College, and the Sanskrit College amongst others. College Street came into being when David Hare, a Scottish watchmaker and philanthropist, set up the Hindu College in order to offer a liberal education to children of the Hindu community. The College was founded in 1817 with 20 scholars and is still in operation today.

The street is an eclectic mix of formal and informal educational and book-related activities with the bamboo and canvas book stalls sharing space with India's largest publishing houses. The local community refer to the street as 'Boi Para' which means book colony and the area is the centre of Kolkata's literary and intellectual life. Buyers can browse the stalls and hundreds of bookshops and, although books are sold very cheaply, it is customary to haggle. On sale are pamphlets, first editions, and paperbacks in all the

Indian languages plus French, German, Russian, and English. Book browsers are welcome to sit and read in the bookshops for as long as they wish without buying (Patra, 2020).

Along with the attraction of the books overflowing in piles onto the pavement, College Street also offers visitors a wide range of food outlets including restaurants, street food, and sweet shops. The famous Coffee House which was founded in 1942 as the Indian Coffee House is a meeting place for writers, students, and intellectuals and has long been a favourite with artists and political leaders. Post-independence, the name changed to the Coffee House, and it was threatened with closure in 1958, but following a petition set up by academics of the nearby Presidency College and the University of Calcutta, it was saved on the basis of its historic building. It now attracts a large clientele of students and tourists. Tour operators offer guided tours of the area, and it features in city guidebooks. Its fame was sealed when it was named in *Time Magazine's* Best of Asia listing in 2007 and this literary district is one of the most famous landmarks in India. The International Kolkata Book Fair also contributes to the city's literary reputation.

College Street did not escape the twin impact of the global Covid-19 pandemic and Cyclone Amphan which hit the city in May 2020. The closure of the nearby academic institutions and lack of tourists and local business meant that book stall owners were forced into setting up online sales and home delivery services. Some owners have started opening their shops on Sundays to try to regain lost revenue (Javed, 2022).

As a form of regional development, book towns are considered to be a sustainable option, re-purposing empty buildings and demanding little in the way of infrastructural development. They can generate income through increased visitor numbers and spend and encourage the development of creative hubs, focussing on writing, illustrating, bookbinding, calligraphy, papermaking, and other associated crafts. The social and economic benefits become even more marked when the book town has an associated literary festival as we shall discover in the following section.

Literary festivals

According to McGuckin (2022, p. 130), literary festivals are 'primarily public events where literary works are read and discussed by authors, experts, critics, publishers, translators and the public who attend these events.' Europe's oldest surviving literary festival is the Cheltenham Literature Festival in the UK which was founded in 1949. There are now over 450 literary festivals across English speaking countries (Rossetti and Quinn, 2021) but the concept is spreading with the Global Association of Literary Festivals reporting a membership that represents most of the world's regions (Global Association of Literary Festivals, 2023). Festivals can bring a range of benefits, both for the author whose work is being featured and promoted, and for the visitor who is offered an engaging and immersive experience which takes the solitary act of reading into a more performative, co-created space (Robertson and Yeoman, 2014). The host community gains the advantage of networks and participation, and a successful festival will create a distinctive image of the host region as a cultural tourism destination.

It is interesting to note the success of the literary festivals which developed alongside the UK's book towns, Wigtown and Hay-on-Wye. Wigtown's festival now generates over £4m a year and sufficient visitors to support the continuing expansion of book related businesses. This has provided employment which has brought families back to the small Scottish town (Wigtown Book Town, 2023). In Hay-on-Wye, the book festival has been so successful that it has been exported around the world with Hay Festivals being held in Mexico, Spain, and Columbia. On its home ground in 2019, in the last event before the Covid-19 pandemic hiatus, the Hay Festival offered 600 events, sold 278,000 tickets, and generated around £25.6m of visitor spend in the local economy (Hay Festival, 2019). These festivals have a generic literary theme, but increasingly more niche festivals are appearing targeting readers of crime fiction, fantasy, or science fiction (McGuckin, 2022).

Literary festivals incorporate live readings, talks, and discussions with authors, book launches and signings, award ceremonies, theatrical performances, tours and trails in the local area, and opportunities to meet authors and like-minded participants. Such activities enhance the prestige of a region which is capable of attracting the creative classes (Robertson and Yeoman, 2014) and bolsters the status of participants; for example, Rossetti and Quinn's (2021) study of rural literary festivals found that attendees could be reinforcing, activating, or acquiring cultural capital through the activities undertaken. Even the souvenir branded book-bags that participants routinely carry around literary festivals are consciously signalling cultural status to their literary peers and conveying a sense of ownership of the festival itself (Dane, 2020).

Conclusion

In this chapter, we have explored the city as a place of literary reputation, unearthing the markers of literariness found on its streets which coalesce to produce a sense of intellectual distinction. These literary signs were found in street names and statues, in cafés and hotels, and in museums, libraries, and bookshops which celebrate individual authors, works, locations, and the more diffuse activities of reading and writing. Further incorporated by initiatives such as UNESCO's Cities of Literature, the International Book Town network, and the increasing profusion of literary festivals, literature is used not just to promote a city's cultural capital to potential visitors but to sustain the production and consumption of literature. In this context literary reputation is understood as 'more than just a birthplace of a writer or place depicted by a great novel – literary heritage itself becomes part of the place narrative and identity, providing the framework to foster and develop creativity' (Marques, 2019, p. 69). For many places, a literary past and present is therefore a valuable asset, a source of pride, and may even be central to the character of the city itself.

In the next chapter we investigate literary tours and trails which both present an opportunity for combining a range of different literary tourism experiences into one meaningful product. Trails and tours also provide a linear experience as opposed to the nodal sites which characterise some of the settings already explored. To follow in the footsteps of your favourite author or travel with a guide through an imaginary setting can be a powerful experience and these literary journeys are a popular means by which literary tourists read the literary landscapes. As we will discover, they are also an effective means by which destinations can interpret their literary heritage.

References

ALVA (2023) Visits made in 2022 to visitor attractions in membership with ALVA. Available at: https://www.alva.org.uk/details.cfm?p=617 (Accessed: 20 July 2023).

American Writers' Museum (2023) American Writers' Museum. Available at: https://americanwritersmuseum.org/about/ (Accessed: 10 February 2023).

Çevik, S. (2022) 'Literary hotels: A new type of literary tourism or just a product', *Dos Algarves: Tourism, Hospitality and Management Journal*, 42, pp. 25–48. Available at: https://doi.org/10.18089/DAMeJ.2022.42.2 (Accessed: 4 March 2023).

Cities of Literature (2023) About the Cities of Literature. Available at: https://www.citiesoflit.com/about-us (Accessed: 21 July 2023).

Connell, P. (2005) 'Death and the author: Westminster Abbey and the meanings of the literary monument', *Eighteenth-Century Studies*, pp. 557–585. Available at: https://www.jstor.org/stable/30053419 (Accessed: 3 July 2023).

Dane, A. (2020) 'Cultural capital as performance: Tote bags and contemporary literary festivals', *Mémoires Du Livre/Studies in Book Culture*, 11 (2), pp. 1–29. Available at: https://doi.org/10.7202/1070270ar (Accessed: 2 July 2023).

Driscoll, B. (2018) 'Local places and cultural distinction: The booktown model', *European Journal of Cultural Studies*, 21 (4), pp. 401–417. Available at: https://doi.org/10.1177/1367549416656856 (Accessed: 21 July 2023).

Edinburgh Museums (2023) The Writers' Museum. Available at: https://www.edinburghmuseums.org.uk/venue/writers-museum (Accessed: 19 June 2023).

Global Association of Literary Festivals (2023) Global Association of Literary Festivals. Available at: https://www.gaolf.org/ (Accessed: 21 July 2023).

Hay Festival (2019) Hay Festival 2019 Report. Available at: https://cloud.hayfestival.com/wales/reports/Hay-Festival-2019-Report.pdf (Accessed: 5 January 2023).

IOBT (2023) International Organisation of Book Towns. Available at: http://www.booktown.net/ (Accessed: 10 July 2023).

Javed, Z. (2022) 'Desperate times call for desperate measures in Kolkata: Boipara opens on Sundays to get over Covid blues', *The Times of India*, 23 May. Available at: https://timesofindia.indiatimes.com/city/kolkata/desperate-times-call-for-desperate-measures-boipara-opens-on-sundays-to-get-over-cov-blues/articleshow/91730141.cms (Accessed: 11 March 2023).

Kostanski, L. and Puzey, G. (2014) 'A tale of two cities of literature: Toponymic identity and the promotion of Edinburgh and Melbourne in the UNESCO Creative Cities Network', *Els noms en la vida quotidiana: Actes del XXIV Congrés Internacional d'ICOS sobre Ciències Onomàstiques*, 6, pp. 1215–1224. Available at: doi:10.2436/15.8040.01.125.

KZN Literary Tourism (2023) KZN Literary Tourism. Available at: https://literarytourism.co.za/ (Accessed: 17 November 2022).

Lainsbury, A. (2019) 'Library tourism', in Jenkins, I. and Lund, K.A. (eds.) *Literary tourism: Theories, practice and case studies*. Wallingford: CABI, pp. 106–117.

Macleod, D. (2009) 'Scottish theme towns: Have new identities enhanced development?', *Journal of Tourism and Cultural Change*, 7 (2), pp. 133–145. Available at: https://doi.org/10.1080/14766820903126460 (Accessed: 21 July 2023).

Marques, L. (2019) 'The making of the literary city: Edinburgh, Barcelona and Obidos', in Jenkins, I. and Lund, K.A. (eds.) *Literary tourism: Theories, practice and case studies*. Wallingford: CABI , pp. 57–70.

McGuckin, M. (2022) 'Literary festivals: Insights from Ireland', *Consumer Behavior in Tourism and Hospitality*, 17 (2), pp. 129–142. Available at: https://doi.org/10.1108/CBTH-05-2021-0129 (Accessed: 30 May 2023).

Mintel (2011) Literary tourism–international–September. London: Mintel International Group.

MoLI (2023) Museum of Literature Ireland. Available at https://moli.ie/ (Accessed: 1 July 2023).

Patra, S. (2020) 'Literary tourism: Kolkata, India', *Bookriot.com*, 11 November. Available at: https://bookriot.com/literary-tourism-kolkata-india/ (Accessed: 12 January 2023).

Ridanpää, J. (2011) 'Pajala as a literary place: In the readings and footsteps of Mikael Niemi', *Journal of Tourism and Cultural Change*, 9 (2), pp. 103–117. Available at: https://doi.org/10.1080/14766825.2011.562979 (Accessed: 21 July 2023).

Robinson, M. and Andersen, H.C. (2002) 'Reading between the lines: Literature and the creation of touristic spaces' in Robinson, M. and Andersen, H.C. (eds.) *Literature and tourism: Essays in the reading and writing of tourism*. London: Thomson, pp. 1–38.

Robertson, M. and Yeoman, I. (2014) 'Signals and signposts of the future: Literary festival consumption in 2050', *Tourism Recreation Research*, 39 (3), pp. 321–342. Available at: https://doi.org/10.1080/02508281.2014.11087004 (Accessed: 21 July 2023).

Rodrigues de Sousa, S. (2022) 'Book shop tourism' in Quinteiro, S. and Marques, M.J. (eds.) *Working definitions in literature and tourism: A research guide*. Faro: University of Algarve, pp. 13–14.

Roque, M.I. and Guerreiro, D. (2021) 'Reading the tourist destination: Bibliotourism and place perception', *Spatial and Organizational Dynamics*, 9 (1), pp. 42–60. Available at: http://hdl.handle.net/10400.14/33186 (Accessed: 21 July 2023).

Rossetti, G. and Quinn, B. (2021) 'Understanding the cultural potential of rural festivals: A conceptual framework of cultural capital development', *Journal of Rural Studies*, 86, pp. 46–53. Available at: https://doi.org/10.1016/j.jrurstud.2021.05.009 (Accessed: 21 July 2023).

Royal Mail (2020) A nation of book worms: Royal Mail reveals house and street names with literary namesakes across the UK. Available at: https://www.poweredbypaf.com/a-nation-of-book-worms-royal-mail-reveals-house-and-street-names-with-literary-namesakes-across-the-uk/ (Accessed: 21 July 2023).

Savage, K. (2007) *History, memory, and monuments: An overview of the scholarly literature on commemoration* Pittsburgh: University of Pittsburgh.

Seaman, A.N. (2019) 'Making literary places', in Jenkins, I. and Lund, K.A. (eds.) *Literary tourism: Theories, practice, and case studies.* Wallingford: CABI, pp. 140–148.

Seaton, A.V. (1999) 'Book towns as tourism developments in peripheral areas', *International Journal of Tourism Research*, 1 (5), pp. 389–399. Available at: https://doi.org/10.1002/(SICI)1522–1970(199909/10)1:5%3C389:AID-JTR204%3E3.0.CO;2–0 (Accessed: 21 July 2023).

Smith, A. (2007) 'Monumentality in "capital" cities and its implications for tourism marketing: The case of Barcelona', *Journal of Travel & Tourism Marketing*, 22 (3–4), pp. 79–93. Available at: https://doi.org/10.1300/J073v22n03_07 (Accessed: 21 July 2023).

Tokić, K. and Tokić, I. (2018) 'Tourism potential of libraries', *Tourism: An International Interdisciplinary Journal*, 66 (4), pp. 443–460. Available at: https://hrcak.srce.hr/214041 (Accessed: 21 July 2023).

Wigtown Book Town (2023) Wigtown Book Town. Available at: https://www.wigtown-booktown.co.uk/ (Accessed: 21 July 2023).

Young, L. (2015) 'Literature, museums, and national identity, Or, why are there so many writers' house museums in Britain?', *Museum History Journal*, 8 (2), pp. 229–246. Available at: https://doi.org/10.1179/1936981615Z.00000000052 (Accessed: 21 July 2023).

8 On the trail of literary fiction

Routes and tours

In the preceding chapters we have explored a range of literary fiction sites, whether they be autobiographical, imaginary, or socially-constructed. We have also considered the role that literature plays in the evolution of cities renowned for their cultural status. In this chapter, we examine the case of literary trails and tours which, in their ambition to package together literary fiction themes into a coherent product, inevitably use many of the attractions and experiences already discussed. What differentiates the trail, route, or tour is the linear nature of the experience where the literary themes are consumed as part of a (self) guided voyage of discovery through the life and works of a literary figure. Very often conceived as 'following in the footsteps of' (Schaff, 2011), these tourism products have become increasingly popular in recent decades although the trope of the literary journey featured heavily in early literary tourism pilgrimages as we saw in Chapter Two.

A literary trail is a self-guided, themed walk (or drive) which includes a range of different sites, monuments or landscapes associated with an author and their works, and which are generally open to be visited by the general public. Trail guides and interpretive material are usually free and often produced by local tourism authorities. A literary tour shares some of these characteristics but is usually a commercial, guided experience, with an expert guide being one of the key benefits, often unlocking 'behind the scenes' access to paying tour members. In both cases, these themed journeys will allow a town or region to direct the tourist gaze (see Chapter Two) and construct itself as a literary destination either following in the footsteps of a single literary luminary or celebrating a group of writers connected by place. The actual journey that is produced encourages literary pilgrims to follow in the tracks of their heroes and the literary theme provides an attractive and advantageous cultural package for guides offering tours.

Literary trails

Trails are 'a means of directing the visitor experience by providing a purposeful, interpreted route' (MacLeod, 2017, p. 423) and these popular tourist products are found in urban, rural, and coastal settings. There are even

DOI: 10.4324/9781003041740-8

examples of underwater diving trails (Wegner, Tonioli and Cabral, 2006) where archaeological sites and marine life are the focus of the route. Of course, humans have been creating and following paths for millennia and these routes have formed the basis of mobility patterns throughout the ages, whether for trade, migration, or pilgrimage. However, 'while they have been recognised elements of human landscapes, the contribution they have brought to tourism and recreation has been understated' (Timothy and Boyd, 2015, p. 1). In particular, the use of a trail as a demarcated route to a final destination can be contrasted with the paths traditionally taken by hunters and traders who would have achieved their goals along the way. Thus, the concept of the interpreted trail focuses on the enjoyment and discovery of the journey itself rather than the final stopping point.

The development of literary trails

The roots of the tourist trail go back to ancient pilgrim and trade routes but it is suggested by Aitchison, MacLeod, and Shaw (2002) that the appreciation of guided walking began in the eighteenth century with the production of a number of guides including the much-used, illustrated *Picturesque Tours of England and Scotland* which 'led visitors on a pre-ordained route through places of literary and artistic association' (p. 79). This concept of 'informed urban walking' (Goodey, 1975, p. 29) aroused interest in conservation bodies from the 1970s onwards with the development of nature trails. Many of the United Kingdom's (UK) walking trails began life as military or drovers' roads and have only become re-configured as leisure experiences within the last 40–50 years. The number and variety of tourist routes and trails has expanded markedly with Hayes and MacLeod (2007) identifying at least 1,300 in England alone at the beginning of the twenty first century. They vary significantly in terms of scale from the international to the local. There are examples of international routes (such as the Council of Europe Cultural Routes which span several member states, see Table 8.1) and national trails, such as the 700km long Slovenian Writers Trail which includes over a hundred different sites (Potočnik Topler, 2016). Regional trails, such as the Robert Burns Trail in Dumfries and Galloway in Scotland will include a number of towns and settlements, and local trails generally explore a handful of sites within a city or village setting such as Hong Kong's Southern District Literary Trail. Literary trails can even be site-based, for example the Literary Trail offered at the University of Oxford's Botanical Garden and Arboretum.

The majority of trails are developed within partnerships, usually led by local authority cultural, leisure, recreation, and tourism departments. Partners in trail development include tourism stakeholders, local development agencies, conservation bodies, civic societies, local history groups, and, in the case of literary trails, trusts and literary estates (Hayes and MacLeod, 2008). Given the nature of these partnerships, it is not surprising to find that their development is usually justified in terms of visitor information, education,

Table 8.1 Council of Europe Literary Cultural Routes

Council of Europe Cultural Routes were first launched in 1987. There are currently 45 Cultural Routes which encourage travel 'to discover the rich and diverse heritage of Europe by bringing people and places together in networks of shared history and heritage.' Themes include art, industry, music, history, and literature (Council of Europe, 2023).

Literary Cultural Route	**Description**
In the Footsteps of Robert Louis Stevenson	Founded in 2015. Crosses France, Germany, and the UK.
European Route d'Artagnan	Founded in 2021. Crosses Belgium, France, Germany, Italy, Netherlands, and Spain.
Women Writers Route	Founded in 2022. Crosses Bulgaria, Croatia, Montenegro, Poland, Serbia, and Slovenia.

and place marketing but as Robinson and Andersen (2002) suggest, trails address more than these three objectives, as 'linking individual attractions and sites of interest together creates a more potent and penetrative tourist product, not just in promotional terms but in a wider intellectual sense too' (p. 22). Ward-Perkins, Beckmann, and Ellis (2019) comment on the structure and efficiency that trails offer tourism suppliers and the sense of adventure that following a themed route gives to the visitor.

One of the reasons that trails have proliferated in recent years is that they are relatively cheap to set up and may require little infrastructural development apart from interpretative materials and applications, thus offering a sustainable option for tourism product development. For a relatively small outlay, they can therefore bring considerable benefits to a destination and to visitors. In terms of visitor management, a set of trails can provide a range of different experiences which encourage longer stays and re-route visitors away from more congested sites. The popular and long-established German Fairytale Route, for example, which leads from Hanau to Bremen, is now 'crisscrossed' with the Half-Timbered Housing Route, the Wiser Renaissance Route, and the Asparagus Route of Lower Saxony. This collection of super-imposed trails 'impregnates the same landscape with diverse overlapping narratives' (Hemme, 2005, p. 75).

By linking the trails with relevant accommodation, catering, and retail businesses, the visitor experience can be augmented and additional spend is generated. Walking trails can offer a sustainable and healthy visitor experience especially when they are integrated into the public transport network with updated and reliable information made available to users. Communities and local economies can also be sustained by the development of trails as the partnerships and networks which are necessary to their inception can provide ongoing support. Additionally, a literary trail can also be valuable in reflecting local identity, especially if its development involves communities and individual stories. The Mississippi Writers Trail, which features in Box 8.1, is a good example of a developing literary trail which invites the local community to make decisions on which writers to include, and which focusses on the state's considerable and proud literary legacy rather than its more difficult history.

Box 8.1 Case study: The Mississippi Writers Trail

The Mississippi Writers Trail was founded in 2018 and 'pays tribute to the state's most acclaimed and influential writers through a series of historical markers that recognize the importance of place in an author's life while educating the public about the history and legacy of Mississippi writers' (MAC, 2023). The Trail was set up by the state's arts grant-giving body, the Mississippi Arts Commission, and funded by the state government and National Endowment for the Arts. Described as a 'statewide community improvement activity' (MAC, 2023), the Trail is designed to promote historical preservation and encourage tourism.

There are currently 14 markers on the trail, including literary greats such as Eudora Welty, William Faulkner, and Tennessee Williams along with a number of contemporary writers such as novelist Jesmyn Ward who won the National Book Award in 2011 and 2017.

The trail markers are made of cast aluminium and are shaped like an open book giving enough space for biographical information and details on literary settings. Novelist Margaret Walker's marker is placed on the campus of Jackson State University where she taught for 30 years. Her marker gives us basic biographical details but also outlines her career, her academic accomplishments, and awards received. We learn that

> after decades of research, she published her 1966 novel *Jubilee* that ushered in a genre of modern slave narratives. At the forefront of the Black Studies and Black Arts Movements, in 1968 she founded an institute at JSU to preserve black history and culture.
>
> (Mississippi Historical Markers, 2023)

For a state known for the role it played in the Civil Rights movement and for its musical legacy, the Writers Trail allows Mississippi's literary heritage to come to the fore. There is a high level of community involvement in the Trail with members of the public being able to nominate specific writers for inclusion, and local communities being involved in determining the location for markers.

Public nominations are reviewed by a committee of scholars against a set of criteria. In particular, writers must have a strong association with the state of Mississippi and have a national reputation. Additionally, they may also be in receipt of a significant award for their writing, have historical significance, and be a strong literary influence on other writers.

Living authors are consulted on the text for their markers and their locations. In an interview in the *New York Times*, writer Richard Ford explained that although the obvious site for his marker would be his childhood home in Jackson (where he attended the same school as Eudora Welty), he wanted his to be located in the Carnegie Public Library in the small town of Clarksdale in the Mississippi Delta, explaining that the library was his haven and he wished to be remembered there, where he wrote his 1986 novel *The Sportswriter* (Holson, 2018).

The 14 markers are scattered throughout the state of Mississippi and more will be included over time. It is intended that this will increase tourism, particularly to small towns which have few alternative visitor attractions.

We have seen that literary trails can focus on individuals or on wider literary reputations and in their study MacLeod, Hayes, and Slater (2009) developed a typology of three different literary trail categories which, to some extent, mirror the different categories of literary tourism in more general terms. Firstly, *biographical trails* focus on the life and times of the author and are aimed at literary pilgrims with pre-existing knowledge. Literary trusts and other custodians have a high level of involvement in the development of this trail type. Secondly, they identified the *literary landscape trail* which utilises 'the association between the landscape and the author or their literary creations as a means of communicating a compelling brand identity for a region' (MacLeod *et al.*, 2009, p. 166). This trail type is often developed by partnerships including local authorities, tourist boards, and the private sector. Finally, the authors identified the *generic literary trail*, developed by local authorities to heighten awareness of the literary reputation of a region with the resulting boost to civic pride and visitor numbers.

Whilst most trails are self-guided, some can be monetised by providing occasional tours with guides and additional experiences such as readings and performances. The 24km long Basinkta Literary Trail in Lebanon incorporates 22 literary landmarks linked to poets and novelists from the region such as Mikhail Naimy, Amin Maalouf, and Abdallah Ghanem and professional tour guides are available to accompany walkers (Lebanon Mountain Trail Association, 2023). Similarly, tourists following the *Shadow of the Wind* Trail in Barcelona which follows the settings of Carlos Ruiz Zafon's novel can book a tour guide who will read appropriate passages at various locations, thus enhancing the visitor experience as recent enthusiastic reviews on Tripadvisor suggest (Tripadvisor, 2023).

Once a trail has been established and is being used, it is important to ensure that it continues to be monitored and upgraded to ensure that signage remains relevant and legible and interpretation content is up to date. However, this type of ongoing commitment is not always built in at the development stage and trails can fall into disrepair due to confusions over which authority has overall responsibility (Hayes and MacLeod, 2008). The Tin Tin

Trail (the only literary trail based on a comic series) which was launched in the town of Saint Nazaire on the west coast of France, incorporated into its plans a mechanism for reporting any damage to the large, coloured metal panels representing Hergé's story of The Seven Crystal Balls (1948), and consequently, the trail is regularly maintained by the local authority (Bréus, 2020).

Trail interpretation

Traditionally, a literary trail (or indeed any themed trail) would be interpreted by means of a hand-held printed map with the literary route superimposed on the existing cartography. These maps might be augmented by plaques affixed to buildings and by interpretive panels along the route. Some locations commit fully to their trails by incorporating quotations from authors carved into the footpaths, for example in the Te Papa Ōtākaro/Avon River Precinct Literary Trail in Christchurch, New Zealand. Despite the transformational effect that digital technology has had in heritage interpretation, the simple trail map can still be found in many settings, for example the Jane Austen Trail in Chawton, Hampshire and the four John Steinbeck routes through the Salinas Valley, California which are available as paper maps and online, as printable PDFs. These simple maps combine places with autobiographical connections to the writers and fictional settings to produce a literary lens through which the visitor gazes. The Steinbeck Trails, for example, include a number of the author's childhood and adult homes, places where he and his parents worked, numerous settings for his novels, and examples of landscapes that were the backdrops to his tales. Inevitably, Steinbeck's grave also features. These examples of simple literary cartography are often augmented by literary biographies and history for example in the Grey Street Trail in Durban or with excerpts from literary works as can be found in the Sarrià Old Town section of the Literary Map of Barcelona which, for example, points out the houses of Gabriel García Márquez, Mario Vargas Llosa, and Joan Vinyoli, interspersing these points of interest with extracts and poems.

However effective the simple literary map may be, Quinteiro (2020) reminds us that the audience for literary tourism uses digital technology in their everyday lives and are accustomed to accessing cultural products through their devices. Therefore, digital applications should be used to interpret literary sites and attract visitors to places of literary association. There are already many examples of the adoption of technology in literary trails interpretation, the earliest being audio trails which incorporate commentary, readings, and local history and have been found to be particularly effective in encouraging visitor engagement with place (Butler, 2007). More recently, readable Quick-Response (QR) codes have been integrated into trail interpretation which, when scanned by a user's mobile phone or tablet, will unlock a wealth of additional interpretive material. The Agatha Christie Trail in Puerto Pollença, Mallorca, part of the Fundació Mallorca Literària's Walking on Words literary mapping project, uses interpretive pillars as markers.

These pillars provide relevant extracts from Christie's work but also incorporate QR codes which allows the visitor to access much additional material, such as readings, maps, photographs, and biographical detail which helps to place the marker within its wider context. This technology is extremely useful as it offers unobtrusive interpretation which can be easily updated remotely.

Mapping technology also clearly lends itself to the development of literary trails, for example the University of Oxford's literary trail around their botanical garden uses three-word mapping to help visitors locate specific plants which feature in literary texts from Philip Pullman's *The Amber Spyglass* (2000) to Evelyn Waugh's *Brideshead Revisited* (1945). The specific mapping details are displayed on a dedicated webpage and accompanied by recorded commentaries. Online mapping with hyperlinks allows visitors to visualise the route for wayfaring purposes but also to see images of sites, read extracts *in situ*, and access additional information. The application of such interpretive devices allows for an approach that can be more tailored to the needs of the individual literary tourist, whether they be a well-versed pilgrim in search of biographical detail or a more generalist visitor who is seeking to link a landscape with a classic text from their childhood (Cunningham and Hinze, 2013). The use of technology in the development and presentation of literary tourism will be discussed further in Chapter Ten.

Figure 8.1 Agatha Christie Trail marker, Puerto Pollença, Mallorca. Photograph by the author.

Literary tours

In the foregoing discussion, we have examined the development and interpretation of self-guided literary routes and trails which create themed journeys that link together places of literary association. In this section, we will turn our attention to literary guided walks and tours which utilise the expertise of the tour guide and the shared cultural capital of the tour group to create the experience, for which the visitor pays. To a much lesser extent, non-commercial walking tours and longer group itineraries are provided by literary societies, writers' groups, and volunteers. Both short, guided walks within a limited locale, and longer tours which may involve several overnight stays and international travel will be considered.

Walking tours

Any city with a claim to a literary heritage will offer commercial guided walking tours, from the plentiful supplies in London, Edinburgh, Dublin, and Paris, to New York, Moscow, Shanghai, Stockholm, Melbourne, Prague, Madrid, Buenos Aires, and many others. These have been a familiar aspect of urban cultural tourism since the 1980s when post-industrial cities began to position themselves through local destination culture aimed at tourists in search of more engaging and fulfilling experiences (see Chapter Two). The walking tour is also the corollary of the rise of the heritage movement and the desire for memorialisation within the urban environment. Thus, walking tours 'conjure up the historical past and map fictional worlds onto the city in order to proffer it as a valuable and memorable experience' (Plate, 2006, p. 110). These tours are usually short, being around two to three hours in length, and they interpret the city through the eyes of one or more authors, or less frequently, through the perspective of a literary figure. The tour guide is usually a local resident which allows for 'insider information' and their literary expertise is at the centre of the experience for the visitors as it is the quality of the guide that provides the interpretation and turns the everyday urban environment into a literary landscape. The guide may be in costume, and it is customary for them to read extracts from texts, sometimes in character.

A fairly standard offering is the three-hour long *Walking with Hemingway: A Journey Through Literary Paris* tour which takes visitors through St Germain in the city's Left Bank. This neighbourhood was the centre of the *avant-garde* movement and housed a wide range of writers and artists who were part of Hemingway's circle. His haunts, homes, and favourite walks are incorporated into the walking tour which is led by a literary scholar. This walking tour explores the life and connections of a literary great within the context of a specific artistic movement. This represents an alternative lens through which to view the tourist city (Context Travel, 2023).

As an experience offered to visitors, the guided walking tour, like a self-guided trail, can provide several different narratives within a single setting, allowing the tour guide to select the landscapes, sites, and stories that best suit their theme and the nature of their audience, be they well-informed enthusiasts, or those with a passing interest. Thus, the walking tour has 'an ability to re-assemble and delineate a particular set of objects, people, discourses, and landscapes, and a flexibility for broad adaptation' (Wang and Kao, 2017, p. 1004).

The practice of walking is suited to the pursuit of literary heritage on both a practical and emotional level. Many historic urban centres, where walking tours usually take place, are inaccessible to vehicles and offer pedestrian zones which allow visitors to get closer to the sites associated with their favourite author. On an emotional level, the act of walking locates the experience especially in tours where visitors are encouraged to follow the author's footsteps. To feel the same cobbles beneath one's feet and admire identical views brings us closer to a cherished author and produces a more embodied experience. Travel at walking speed allows for a more considered engagement with the environment and thus a guided walking tour allows us to make sense of the city, producing a narrative out of the sensory overload that international travel often creates (Plate, 2006).

To travel slowly and mindfully on foot through the city invokes the figure of the *flâneur* who is both at one with the urban scene, but also engages in its critique and its memorialisation. The nineteenth century notion of *flânerie* can still be witnessed in contemporary cities as a specific mode of urban exploration and 'as a particular type of reflective relationship with people and spaces' (Nuvolati, 2014, p. 21). The *flâneur* interprets the present through a historical lens, reading the cityscape. It is a stance often taken by writers who incorporate their musings into their work. Hence, as Plate (2006) suggests, 'the *flâneur* is the reader who takes the knowledge gained from the writings of earlier *flâneurs* to the streets; and *flânerie* is an experience of the urban environment that is mediated by literature' (p. 109). The walking tour can certainly promote this mode of embodied and intellectual engagement and sets this experience very decidedly apart from a coach tour where the sites are seen at a distance and at speed (Reijnders, 2011). The performative nature of walking tours is evidenced through the bodily apprehension of the literary works, whether that be in the sensory experience of the sights, sounds, and smells of the streets or in the more traditional sense of performance (see Box 8.2). It has been noted above that tour guides are sometimes costumed and perform readings of texts. In some walking tours, the participants also take on the guise of author or character and act out key scenes, for example, Reijnders describes a participant of a *Dracula* walking tour of Whitby, in North Yorkshire, re-enacting the nocturnal walk of Lucy Westenra from her home to the graveyard and, ultimately, the Count, complete with white nightgown and streaming hair (Reijnders, 2011).

Box 8.2 Case study: Walking in literary London

London has long been considered a city replete with literary associations which intersect and overlap and free the literary tourist from the 'dictatorial logic of any single author, text, oeuvre, or genre, releasing them into a promiscuously sociable saunter through a canonical litter of biographical anecdote and imaginary episode strewing the streets of the city' (Watson, 2009, p. 139). Consequently, there is a history of suggested walking routes through the capital's morass of literary connection, from Allbut's 1886 *London Rambles 'En Zig-Zag' with Charles Dickens* (Watson, 2009, p. 141) to the various operators offering guided walks to contemporary tourists. Dickens still features heavily in these itineraries, along with Shakespeare and Holmes, but this case study offers an account of a London walking tour exploring T.S. Eliot's poem *The Waste Land*, first published in 1922.

The two-hour long tour is part of an extensive collection of literary London walks offered by a long-established walking tour company. The tour was undertaken by 11 participants – eight women and three men. Ages ranged from young adult to senior citizen and the group comprised mainly American and British citizens. Motivations to join the tour included interest in the poem, a desire to do something different in one's home city, a general interest in London's arts, and a visitor who was a dedicated student of the poet.

The tour guide introduced himself, his career working in the cultural industry, and his expertise as a well-researched, lifelong fan of *The Waste Land*. The tour consisted of 13 stops, the majority being on main streets, one in a courtyard, one in a garden, and two inside churches. The tour was skilfully compiled to include specific London locations mentioned in the poem – these included streets, buildings, viewpoints, and the key feature running through the poem, the river Thames. Geographic linkages were pointed out with readings of appropriate verses to illustrate the merging of the poetic with the topographical. Sites were visited which linked to both the poem and the poet's biography; for example, the Church of Magnus Martyr, whose white and gold Ionian columns are described, was also a favourite place of worship for Eliot. The guide also created a sensory encounter, ensuring that the sights and sounds of the poem were experienced, for example we stopped by a public bar in Lower Thames Street which was still exuding the same 'clatter' and 'chatter' as that depicted in the poem.

Along with literary and biographical sites, the tour included stops which provided supplementary details informing the understanding of the poem or creating an appropriate atmosphere. Thus, participants paused at a statue commemorating Shakespeare's work and here the group considered the poem's numerous references to the Bard. Stopping in a quiet garden off a busy street, the guide discussed the poem's horticultural elements, using the peaceful surroundings to impart some recent scholarship which has added to Eliot's biography. This account provides an example of how the many imaginative, biographical, and sensory elements of the city can be combined into an engaging literary tourism experience through the construction of a creative walking tour.

Literary group tours

As Robinson and Andersen note, ‘the tourism industry is now well versed in structuring the sites and symbols of literature into organized routes’ (2002, p. 22) and another example of this is the complete packaged tour where transport, accommodation, and a literary tour guide are provided. This type of tour is offered by literary societies and writers’ groups where acts of pilgrimage or attempts at authorial inspiration are the underlying motivation but, increasingly, specialist (and more generic) commercial tour operators are now offering these experiences as lucrative and distinctive products.

A commercial literary group tour will certainly incorporate transport, four or five star rated accommodation, tours of specific sites, talks by experts, and the services of a well-qualified guide. However, they may also offer overnight stays in hotels associated with the literary theme, including commemorative dinners and special events, for example, several Dracula-themed tours in Romania include accommodation at the Dracula Castle Hotel, vampire dinners, annual Hallowe’en balls, and even a Ritual Killing of the Living Dead to help a haunted spirit to gain peace (Transylvania Live, 2023). One of the unique benefits of such tours for visitors is the opportunity to access behind-the-scenes experiences not on offer to the general tourist. *Literary Tours of Egypt* is an Egypt-based specialist operator who offers tours of the Nile, Cairo, Luxor, Aswan, and Tell al-Amana, following in the footsteps of writers Agatha Christie, Rosie Thomas, and Noel Barber. The tours are from five to fifteen days in length and the operator also offers a six-day crime writing retreat on board a traditional Nile sailing boat. *Literary Traveler*, an American operator, runs a three-day Literary Tour of Concord, Massachusetts, a well-known literary destination which includes Alcott, Thoreau, and Emerson sites plus talks and outdoor activities. Dublin-based *Celtic Luxury Tours* is a more generalist operator offering tailored tours in England, Scotland, Ireland, and Wales. Themes include golf, castles, spa retreats, genealogy, and gastronomy along with literary tours.

Prices for these experiences are typically high, reflecting the level of personalisation and service on offer and the proposition is generally elitist in nature, making full use of the traveller vs. tourist dichotomy (see Chapter Three). Tour descriptions are replete with language which differentiates their potential clients who are offered ‘curated’ experiences away from the usual ‘tourist traps’ where their knowledge and cultural discernment are recognised. These tours are aimed at the ‘culturally curious’ and ‘intentional traveller’ rather than the mere day-tripper and the ‘golden age of travel’ is frequently invoked. Tour guides are routinely referred to as ‘experts,’ ‘scholars,’ ‘historians,’ and ‘specialists.’ Clearly these tours are high in cultural capital, much of which resides in the expertise of the tour guide which is central to the authenticity and value of the experience (Carvalho, Baptista and Costa, 2012).

Several researchers have explored this in relation to literary tours, for example Douglas Pocock’s (1992) study of participants on a Catherine

Cookson Tour in South Tyneside in the Northeast of England and Earl's (2008) examination of visitors on an Arthurian Tour in Cornwall, Southwest England. In the case of the redeveloped urban settings of the Catherine Cookson Tour, the tour guide's knowledge was essential to help visitors to imaginatively recreate lost literary landscapes. In the example of the Arthurian Tour, highly qualified specialists were prized in that they allowed participants to differentiate themselves from the holidaymaking hordes in their search for authentic traces of the real King Arthur. Current approaches to the role of the tour guide suggest that this figure is no longer merely a presenter and entertainer of their group, engaged in one-way communication, but is, within our new era of more engaged, interactive, and co-created tourism, an experience broker (Weiler and Black, 2015). As such, tour guides mediate the provision of physical access to backstage spaces, broker encounters with host communities and local experts, and provide understanding and empathy to allow the visitor to apprehend the culture of the visited. In providing safe enclaves within which visitors can immerse themselves, a tour guide allows tourists 'to try new experiences while feeling supported' (Parsons, Houge MacKenzie and Filep, 2019, p. 3).

Although a highly qualified tour guide is critical to the success of a literary group tour, the other participants also play a key role. Tours are often advertised as offering 'like-minded' fellow travellers and the level of interaction between participants is important in the creation of the experience, whether it is engaging in discussion, sharing viewpoints, or getting involved in the more performative aspects as outlined above. Being part of a select band of discerning travellers may be a large part of the appeal of a literary tour. As many tour group participants have high levels of knowledge of their favourite author, tours can be conceptualised as literary pilgrimages with all the associated references to shrines, relics, and worship. In this situation, participants may become a *communitas*, a classless society who may share intimacies and, ultimately, transcendence within a liminal space, far away from everyday concerns (see Chapter Three). Of course, in reality, many such tour participant lists will also include friends and partners who have been brought along in varying degrees of willingness and, as Pocock discovered in South Tyneside, may fail to bring with them eyes primed for the literary gaze (1992, p. 240).

Conclusion

We have seen in the discussion above that literary themes lend themselves well to the creation of a wide range of self-guided and guided tours which offer flexible and engaging experiences for visitors and can contribute to the literary capital of the cities and regions within which they are developed. As linear visitor attractions, literary trails and tours constitute their environments, creating new tourist spaces using themes and stories chosen for that purpose. Thus, the trail identifies a sequence of sites to be visited, creating a chronological narrative. Places can therefore be read like texts, producing meaning and coherence. Consequently, the trail narrative produces a type of literary topography which directs the gaze and manages the experience (MacLeod, 2017) – the visitor follows real and virtual markers in the landscapes which combine with their own imaginative mapping and tour guide prompts to produce tourist spaces which are 'increasingly seen as both physically, culturally and symbolically constructed' (p. 424).

Of course, this process of selection and construction attracts critical attention. Firstly, there is the accusation of inauthenticity where a trail presents a chronological narrative 'in an environment which may in fact have had a far less seamless coexistence with the writer' (Stiebel, 2010, p. 6). Secondly, we find critiques of theming and commodification wherein culture is debased by such processes which package it for the marketplace – and in the case of literary tours, aspects of local and national culture can indeed be sold for relatively high prices. Conversely, such tourism products, based as they are on aspects of national culture, may also be criticised for their elitism. However, the flexible nature of trails and tours means that they can be designed not only to bolster the dominant culture but to allow less well-known histories to emerge. Thus, we may obtain 'a recognition of emerging stories rather than a claim to a monolithic identity' (Carson *et al.*, 2017, p. 381). An example of this is seen in Box 8.1 where Mississippi's established heritages of human rights and music have been augmented by a trail celebrating its considerable literary legacy. Of course, care must be taken when developing literary tourist products, whether they be self-guided trails or group tours, which showcase cultures associated with less affluent communities. The provision of 'slum tourism' where well-heeled city visitors are taken to explore the culture of poorer neighbourhoods such as New York's Harlem, Los Angeles' Compton, and the inner city of Johannesburg, needs to be carefully managed and ethical issues around consent and exploitation must be considered (Giddy and Hoogendoorn, 2018).

This chapter has focussed on the synergies between literature and geographical space in the creation of literary topographies. In the following chapter, we will explore further alliances between literature and related fields. In doing so, we will explore the emerging field of contents tourism which will embrace the synergies between literary and media-based tourism, literary performance, and merchandising.

References

Aitchison, C., MacLeod, N. and Shaw, S. (2002) *Leisure and tourism landscapes: Social and cultural geographies*, Abingdon: Routledge.

Bréus, S. (2020) '*Comics-based literary trails: The case of Tintin and the Saint-Nazaire landmarks*', PhD thesis, University of Plymouth, Plymouth. Available at: https://dx.doi.org/10.24382/683 (Accessed: 4 April 2023).

Butler, T. (2007) 'Memoryscape: How audio walks can deepen our sense of place by integrating art, oral history and cultural geography', *Geography Compass*, 1 (3), pp. 360–372. Available at: https://doi.org/10.1111/j.1749-8198.2007.00017.x (Accessed: 18 November 2022).

Carson, S., Hawkes, L., Gislason, K. and Cantrell, K. (2017) 'Literature, tourism and the city: Writing and cultural change', *Journal of Tourism and Cultural Change*, 15 (4), pp. 380–392. Available at: https://doi.org/10.1080/14766825.2016.1165237 (Accessed: 3 December 2022).

Carvalho, I., Baptista, M.M. and Costa, C. (2012) 'Binding culture and tourism through literature: The potential of literary tourism', *International Journal of Hospitality and Tourism*, 2 (2), pp. 1–27.

Context Travel (2023) Walking with Hemingway: A journey through literary Paris. Available at: https://www.contexttravel.com/cities/paris/tours/paris-literature-tour (Accessed: 2 April 2023).

Council of Europe (2003) Cultural routes. Available at: https://www.coe.int/en/web/cultural-routes/all-cultural-routes (Accessed: 15 May 2022).

Cunningham, S.J. and Hinze, A. (2013) 'Supporting the reader in the wild: Identifying design features for a literary tourism application', in *Proceedings of the 14th Annual ACM SIGCHI_NZ conference on Computer-Human Interaction*, pp. 1–9. Available at: https://doi.org/10.1145/2542242.2542253 (Accessed: 2 June 2023).

Earl, B. (2008) 'Literary tourism: Constructions of value, celebrity and distinction', *International Journal of Cultural Studies*, 11 (4), pp. 401–417. doi:10.1177/136787790809600.

Giddy, J.K. and Hoogendoorn, G. (2018) 'Ethical concerns around inner city walking tours', *Urban Geography*, 39 (9), pp. 1293–1299. Available at: https://doi.org/10.1080/02723638.2018.1446884 (Accessed: 5 May 2023).

Goodey, B. (1975) 'Urban trails: Origins and opportunities', *The Planner*, 61, pp. 29–30.

Hayes, D. and MacLeod, N. (2007) 'Packaging places: Designing heritage trails using an experience economy perspective to maximize visitor engagement', *Journal of Vacation Marketing*, 13 (1), pp. 45–58. doi:10.1177/1356766706071205.

Hayes, D. and MacLeod, N. (2008) 'Putting down routes: An examination of local government cultural policy shaping the development of heritage trails', *Managing Leisure*, 13 (2), pp. 57–73. Available at: https://doi.org/10.1080/13606710801933420 (Accessed: 25 February 2023).

Hemme, D. (2005) 'Landscape, fairies and identity: Experience on the backstage of the fairy tale route', *Journal of Tourism and Cultural Change*, 3 (2), pp. 71–87. Available at: https://doi.org/10.1080/09669580508668488 (Accessed: 29 May 2023).

Holson, L.M. (2018) 'Marking Mississippi's Literary Trail, from William Faulkner to Jesmyn Ward', *The New York Times*, April 27. Available at: https://www.nytimes.com/2018/04/27/books/books-mississippi-faulkner-trail.html (Accessed: 15 May 2023).

Lebanon Mountain Trail Association (2023) The Lebanon Mountain Trail. Available at: https://www.lebanontrail.org/home (Accessed: 10 July 2023).

MAC (2023) Mississippi Writers Trail. Available at: https://arts.ms.gov/writers-trail/ (Accessed: 3 April 2023).

MacLeod, N. (2017) 'The role of trails in the creation of tourist space', *Journal of Heritage Tourism*, 12 (5), pp. 423–430. Available at: https://doi.org/10.1080/1743873X.2016.1242590 (Accessed: 25 February 2023).

MacLeod, N., Hayes, D. and Slater, A. (2009) 'Reading the landscape: The development of a typology of literary trails that incorporate an experiential design perspective', *Journal of Hospitality Marketing & Management*, 18 (2–3), pp. 154–172. Available at: https://doi.org/10.1080/19368620802590183 (Accessed: 25 February 2023).

Mississippi Historical Markers (2023) Mississippi historical markers. Available at: https://www.mississippimarkers.com/ (Accesssed 11 February 2023).

Nuvolati, G. (2014) 'The flâneur: A way of walking, exploring and interpreting the city', in Shortell, T. and Brown, E. (eds.) *Walking in the European city: Quotidian mobility and urban ethnography*. Abingdon: Routledge, pp. 21–40.

Parsons, H., Houge Mackenzie, S. and Filep, S. (2019) 'Facilitating self-development: How tour guides broker spiritual tourist experiences', *Tourism Recreation Research*, 44 (2), pp. 141–152. Available at: https://doi.org/10.1080/02508281.2019.1582159 (Accessed: 29 May 2023).

Plate, L. (2006) 'Walking in Virginia Woolf's footsteps: Performing cultural memory', *European Journal of Cultural Studies*, 9 (1), pp. 101–120. doi:10.1177/1367549406060810.

Pocock, D. (1992) 'Catherine Cookson country: Tourist expectation and experience', *Geography*, pp. 236–243. Available at: https://www.jstor.org/stable/40572194 (Accessed: 3 September 2022).

Potočnik Topler, J. (2016) 'Literary tourism in Slovenia: The case of the Prežihov Voranc cottage', *Informatologia*, 49 (3–4), pp. 129–137. Available at: https://hrcak.srce.hr/173839 (Accessed: 2 February 2023).

Quinteiro, S. (2020) 'Link up with technology application in literary tourism', in Hassan, A. and Sharma, A. (eds.) *The Emerald handbook of ICT in tourism and hospitality*. Bingley: Emerald Publishing Limited, pp. 379–389.

Reijnders, S. (2011) 'Stalking the Count: Dracula, fandom and tourism', *Annals of Tourism Research*, 38 (1), pp. 231–248. Available at: https://doi.org/10.1016/j.annals.2010.08.006 (Accessed: 18 April 2023).

Robinson, M. and Andersen, H.C. (2002) 'Reading between the lines: Literature and the creation of touristic spaces' in Robinson, M. and Andersen, H.C. (eds.) *Literature and tourism: Essays in the reading and writing of tourism*. London: Thomson, pp. 1–38.

Schaff, B. (2011) '"In the footsteps of…" The Semiotics of literary tourism', *Kultur-Poetik*, 11 (2), pp. 166–180. Available at: https://doi.org/10.13109/kult.2011.11.2.166 (Accessed: 12 April 2023).

Stiebel, L. (2010) 'Last stop "little Gujarat": Tracking South African Indian writers on the grey street writers' trail in Durban', *Current Writing: Text and Reception in Southern Africa*, 22 (1), pp. 1–20. Available at: https://doi.org/10.1080/1013929X2010.9678331 (Accessed: 16 April 2023).

Timothy, D.J. and Boyd, S.W. (2015) *Tourism and trails: Cultural, ecological and management issues*. Bristol: Channel View Publications.

Transylvania Live (2023) Romania Dracula Tour. Available at: https://dracula-tour.com/package-holidays-to-romania/romania-dracula-tour-7-days-dracula-tour-from-bucharest/ (Accessed: 10 April 2023).

Tripadvisor (2023) Shadow of the Wind novel walking tour of Barcelona. Available at: https://www.tripadvisor.co.uk/AttractionProductReview-g187497-d11452082-The_Shadow_of_the_Wind_Novel_Walking_Tour_in_Barcelona-Barcelona_Catalonia.html (Accessed: 8 April 2023).

Wang, C.H. and Kao, Y.T. (2017) 'Re-assembling the memorial landscape: The politics of walking tours in Taipei', *International Journal of Heritage Studies*, 23 (10), pp. 1002–1016. Available at: https://doi.org/10.1080/13527258.2017.1362575 (Accessed: 20 April 2023).

Ward-Perkins, D., Beckmann, C. and Ellis, J. (2019) *Tourism routes and trails: Theory and practice*. Wallingford: CABI.

Watson, N.J. (2009) 'Rambles in literary London', in Watson, N.J. (ed.) *Literary tourism and nineteenth-century culture.* Basingstoke: Palgrave Macmillan, pp. 139–149.

Wegner, E., Tonioli, F.C. and Cabral, D.Q. (2006) 'Underwater trails: A new possibility of marine tourism', *Journal of Coastal Research*, pp. 990–993. Available at: https://www.jstor.org/stable/25741727 (Accessed: 4 October 2022).

Weiler, B. and Black, R. (2015) 'The changing face of the tour guide: One-way communicator to choreographer to co-creator of the tourist experience', *Tourism Recreation Research*, 40 (3), pp. 364–378. Available at: https://doi.org/10.1080/02508281.2015.1083742 (Accessed: 20 April 2023).

9 Literary fiction tourism synergies

The preceding chapters have focussed on an analysis of tourism activity within specific literary locales, whether they be writers' homes, imaginative landscapes, literary theme parks, or walking tours. In these contexts, there is a direct link between geographical space, the literary imagination, authorial biography, and the visitor who is turning an internal narrative into an engagement with real locations. Of course, this is the essence of literary tourism – locating the fictive or chasing the author's shadow in the real world – but in this chapter we will explore some of the wider synergies associated with this activity which augment the experience and extend the range of contexts within which literary tourism is produced and consumed. One of these areas of synergy is the wider media of which literature is a part. In particular, we will explore the relationship between tourism inspired by literature and that induced by film and television programming. Film and TV adaptations are a popular means of consuming literature and viewing these can add to the literary tourism experience by priming the visitor for their encounter. Additionally, locations for popular film and TV adaptations become visitor attractions in their own right, luring media tourists who may or may not have a literary awareness of the site. Literary characters and themes can also travel even further away from the authentic literary locations in the shape of merchandising, where literary characters or artefacts are reproduced on clothing, as toys or games, and as more upmarket literary souvenirs. Characters and events from literature can also be experienced through gamification, performances, literary meals, and special events which may also include dressing up or 'cosplay.' In recent years, tourism which is inspired by various strands of narrative and characterisation has become known in Japan and further afield as 'contents tourism' and this chapter will end with a consideration of this new field and how it allows us to understand the synergies between literary and media-based tourism and other associated practice and consumption.

DOI: 10.4324/9781003041740-9

Screen-based tourism

Film and TV tourism

Print media encourages visitation and promotes particular destinations, as do images that potential visitors consume on the small or big screen. One of the earliest examples of this type of film-induced tourism is the film *Mutiny on the Bounty* (1935) which encouraged wide-scale visitation to Tahiti. A similar effect was experienced after the release of *The Third Man* in 1949 which stimulated tourist interest in Vienna (Roesch, 2009, p. 9). In her seminal work on film-induced tourism, Sue Beeton (2005) states that it is understood 'that the term "movie-induced tourism" relates to on-location tourism that follows the success of a movie made (or set) in a particular region. By using the term "film-induced tourism", this can be expanded to include Television, video and DVD' (p. 9). In our current era, this definition will also include content provided on streaming services. Beeton differentiates between 'on-location' screen tourism where visitors are attracted to a screened location (for example Dubrovnik in Croatia, where *Game of Thrones* was filmed), and 'off-location' screen tourism which includes film studio tours, such as the tour of long-lived British soap opera, *Coronation Street* at the ITV Studios in Manchester. She also categorises 'commercial film tourism' which comprises theme parks and attractions, for example the Wizarding World of Harry Potter attraction at Universal Studios, Orlando, Florida, 'one-off events' such as movie premieres and film festivals, and finally 'sites of mistaken identity' where visitors are attracted to places where a film may be set but is not actually filmed, the most oft-cited example being the film *Braveheart* (1995) which attracted many additional visitors to Scotland but was in fact largely filmed in Ireland.

In 1996 the British Tourist Authority released their Movie Map of Great Britain which capitalised in particular on the many appealing destinations showcased in the film *Four Weddings and a Funeral* (1994). Since then, other national tourist boards have followed suit with dedicated research and guidance produced on maximising the benefits of this type of tourism, for example the *Set in Scotland* film guide produced by Visit Scotland provides information on over 130 Scottish film locations, with the earliest film featured being *The Thirty-Nine Steps* from 1935 (VisitScotland, 2022a). This is an important market for the country with an estimated 656,000 overnight screen tourists visiting Scotland in 2019, spending £64.9 directly on screen-related tourism activities (Screen Scotland, 2021). The recent *Outlander Effect* report (VisitScotland, 2022b) shows the impact of a very popular individual television series on the Scottish economy, as does the *Making of a Screen Tourism Destination* report produced by Tourism Northern Ireland which showcases *Game of Thrones* tourism initiatives in the region (Tourism Northern Ireland, 2021). It is argued that long-running television series may have a deeper impact than films as the 'regular viewing over an extended period of time required by television serial fiction encourages identification, empathy, emotional engagement and parasocial interactions, which in turn motivate spectators to visit the filming locations' (Gómez-Morales, Nieto-Ferrando and Sánchez-Castillo, 2022, p. 73) as evidenced in the examples above. However, although there is potential for tourism product development, destination marketing, and enhanced cultural capital, a link with a very long-running series can also cause problems for the local community as was experienced in the small town of Port Isaac in Cornwall in the West of England which hosted ten series of the comedy drama *Doc Martin* between 2004 and 2022. The series brought much additional tourism, but some residents welcomed the final episode, claiming that the series had turned their home into a circus (Morris, 2020). Even though television series are considered to be the most effective in generating screen tourism there are of course many examples of films which have inspired tourist visits and continue to encourage tourism for several decades. *The Sound of Music* (1965) is one such example, with its blend of nostalgia, soaring musical score, and the prime role played by the Austrian alps and picturesque town of Salzburg securing its longevity.

Screen and literary tourism

A number of the films and television series mentioned above have, of course, been based on successful novels and there is a very strong link between screen and literary tourism as both are motivated by a desire to see, in real life, places which we have virtually visited either through our imaginations or through the media of film (for example see Table 9.1 for a list of Scottish screen locations for films based on novels). In some very well-known cases, for example Harry Potter tourism, where visitors may have consumed both an entire set of novels and film series, it may be almost impossible to untangle the literary from the screen-based motivations. In their work on the relationships between literary, film, and heritage tourism, Agarwal and Shaw (2017) describe the nexus between these related form of tourism as being based on a shared use and commodification of the built, natural, and living heritage, the inter-relationship of the visitor with the cultural resource, and an emphasis on co-creation to produce the heritage, screen, and literary tourism experience. O'Connor and Kim (2014) suggest that film and literary tourism are similar in that they both create new touristic places. However, they differ in the fact that literary tourism is often prompted by biographical interest in a writer, whereas a screenwriter or producer is rarely a focus of tourism interest, with the attention instead placed on the characters portrayed or the actors representing them.

Literary tourists may be interested in augmenting their reading experience by watching film adaptations and visiting a screen tourism location, for example fans of the work of both Lawrence and Gerald Durrell can visit Corfu which offers several biographical sites and experiences associated with the brothers and a range of screen locations linked to the recent television series depicting their lives on the island (see Box 9.1). This could be seen as another example of the concretising power of location tourism (MacLeod,

Table 9.1 Scottish literary screen locations based on novels

Film and literary inspiration	*Scottish locations used in film*
The Thirty-Nine Steps (1935/1959) based on the (1915) novel by John Buchan	• Forth Bridge, near Edinburgh
Ring of Bright Water (1969) based on the (1960) novel by Gavin Maxwell	• Oban, Argyll
The Prime of Miss Jean Brodie (1969) based on the (1961) novel by Muriel Spark	• Edinburgh • Barnbougle Castle, South Queensferry
Kidnapped (1971) based on the (1886) novel by Robert Louis Stevenson	• Stirling Castle, Stirling
Trainspotting (1996) based on the (1993) novel by Irvine Welsh	• Edinburgh • Corrour Railway Station, Highlands • Rouken Glen Park, East Renfrewshire
The Da Vinci Code (2006) based on the (2003) novel by Dan Brown	• Rosslyn Chapel, Midlothian
Cloud Atlas (2012) based on the (2004) novel by David Mitchell	• Scott Monument in Edinburgh • Clackmannanshire Bridge, Fife • Bo'ness and Kinneil Railway, Falkirk • Glasgow
Under the Skin (2013) based on the (2000) novel by Michel Faber	• Auchmithie, Angus • Tantallon Castle, North Berwick • Buchanan Galleries, Glasgow • University of Glasgow
Sunset Song (2015) based on the (1932) novel by Lewis Grassic Gibbon	• Glen Tanar Estate, Royal Deeside
BFG (2016) based on the (1982) novel by Roald Dahl	• Old Man of Storr, Isle of Skye • Old Man of Hoy, Orkney

Source: Adapted from Visit Scotland (2022a) *Set in Scotland: A journey through big screen locations*

Shelley and Morrison, 2018) where visitors seek to anchor their imaginings to a real place. A carefully chosen location which matches the tone and historic period of a literary text will add to the authenticity of this realisation, whether this is experienced by armchair or actual travellers. However, true verisimilitude may be not always be necessary, as we may glean from adaptation theory which arose from the marriage of film studies and literary criticism. Whilst this theory may have begun with a focus on fidelity when translating a story or character from one medium to another, it is now a term 'that has been continuously adapted and it can designate a wide variety of acts of transfer, translation and revision' (Gould and Mitchell, 2010, p. 148). The recent trend for loose film adaptations of literary classics attests to this, for example the 2023 British Broadcasting Corporation's (BBC) adaptation of Dicken's novel *Great Expectations* (1861) which attracted mixed reviews for its modernisation of language and characterisation.

There are many who will watch a film or TV series and visit the location without having read the original and so will remain unaffected by the faithfulness of the adaptation. Others, who may be inspired by a film to go to the original source, will merely have the filmed version as their reference point, for example viewers of Nicholas Cage in *Captain Corelli's Mandolin* (2001) who subsequently contributed to an upturn in sales of the novel of the same name by Louis de Bernières (Busby and O'Neill, 2006). Those who visited the Greek island of Kefalonia in homage to the film or the book would have found actual locations referenced in both media, and hotels and resorts on the island still utilise the association.

Box 9.1 Case study: Locating the Durrells in Corfu

The Ionian Island of Corfu is a good example of a destination with a combined literary and screen tourism offering. For many visitors, the island is closely associated with the Durrell family who moved from England to Corfu in 1935. The family lived in a series of locations across the island and Gerald (author and notable naturalist) and Lawrence (renowned novelist) reflected their experiences in their works. Gerald Durrell's *Corfu Trilogy* is an autobiography recounting the flora and fauna of the island interspersed with the family's exploits. *My Family and Other Animals* (1956), *Birds, Beasts and Relatives* (1969), and *The Garden of the Gods* (1978) sold millions of copies and have been translated into 31 languages. Lawrence Durrell's homage to the island, *Prospero's Cell* was also influential and was published in 1945 to critical acclaim. These literary associations have been one of the island's narratives for many years and bronze relief busts of both brothers are to be found in the Boschetto Gardens overlooking the sea in Corfu Town (see Fig. 9.1). This garden was renamed Durrell Gardens in 2006. However, it was the release in 2016 of a very popular British television series *The Durrells,* dramatizing the lives of the eccentric family in beautiful Corfiot locations, that brought this already popular island to the attention of a wider audience. *The Durrells* ran for four series on ITV until 2019.

The island now fully exploits its Durrell connections with a range of literary and film locations and commercial tours and experiences. There is no attempt at differentiating the media in these products.

Literary tourism sites include:

- Bronze busts in Boschetto Gardens, Corfu Town
- Villa Agazini, the 'Strawberry Pink Villa' in Gerald's books, is set above Perama, near Corfu Town. This house is in private ownership but can be rented as a holiday home.

- The second villa where the Durrells lived was Villa Annemoyani or the 'Daffodil Yellow Villa,' in Gouvia, also close to Corfu Town. This villa is also available for rental.
- The White House, Lawrence and Nancy Durrell's home, still stands at the head of Kalami Bay. Now a café, bar and taverna, it is open to visitors.

Screen tourism sites include:

- The atmospheric Posillipo Villa on the seafront in Kontokali which is the most significant filming location as this house represented the Durrell's villa in the television series. Again, the house, now known as 'The House of the Durrells,' is a holiday rental.
- The replica village of Danilia, modelled on a traditional 1930s Corfiot village which was destroyed in an earthquake, is used as a heritage park, venue, and film set, most notably for the 12th James Bond film *For Your Eyes Only*, and *The Durrells*. Now owned by the Grecotel chain, it can be visited by the public.

A number of Durrells-themed commercial tours are available on Corfu including:

- A three-hour walking tour of Corfu town
- Four-hour coach tours
- A seven-hour boat tour
- Four-hour long shore excursions for cruise passengers
- A six-hour evening tour with dinner
- A ten-day Durrells walking holiday

Screen adaptations of literary texts are an important extension of the works, widening their reach and augmenting the experience of readers. Such adaptations lead to tourism visits from literary, screen, and general visitors and the synergies between the two media reflect the co-creation which exists between reader/visitor and literary representations such that 'while the novel is the direct expression of a writer's interpretation of the fictional world which the writer would like to present to readers, the adaptation may include additional multiple layers of interpretation, transformation, or changes' (O'Connor and Kim, 2014, p. 5). Another increasingly significant layer of interpretation comes in the use of literary texts in the development of computer games as we shall see in the following section.

Computer games and literary tourism

Computer games are games that are played on electronic devices which include specific gaming consoles, smartphones, tablets, or personal computers. Games which include virtual reality also require headsets. There is a broad

Figure 9.1 Bronze bust of Gerald Durrell in Corfu Town, Corfu. Photograph by the author.

range of complexity from simple puzzles for single users to entire virtual worlds with multiple players. Many of these games have sophisticated narrative and character development and writers are employed to develop new story lines and to adapt existing texts to this context. Games which are particularly engaging are described as 'immersive' which means that they use specific techniques to transport the player into an alternative world where they can take on specific personae and undertake tasks whilst following a well-developed story which is conveyed through highly realistic graphics. Several literary texts have undergone this gamification and it is unsurprising that those which involve either detection or voyaging seem to be the most easily translatable. The tales of Sherlock Holmes have been turned into numerous computer games as have the novels of Agatha Christie, for example 'The 4:50 From Paddington' is a hidden object game where the player assumes the persona of Miss Marple. The game is arranged into 12 chapters and players must collect evidence, examine clues, and question witnesses in their attempt to find the killer (The Gamer, 2023). The epic voyages and transformations contained within *The Lord of the Rings* trilogy, *The Hobbit, Treasure Island, Frankenstein*, and *Dracula* have also rendered them easily transposed into the gaming media. The most recent literary-inspired game to be launched is 'Hogwarts Legacy,' published by Warner Bros. This immersive role-playing game sold over 12 million copies and earned around $850 million in its first two weeks (Jacobs, 2023). Its slogan, 'live the unwritten' refers to the fact that the world created in this game predates that of Harry Potter and his friends, being set in the 1800s. Players take on the character of a Hogwarts student who holds the key to a secret that threatens the wizarding world. They control the action as they travel through spectacular interiors and fly broomsticks over aerial views of Hogwarts and its setting. As a virtual experience which stretches the already extensive Potter universe back in time and transports the gamer/traveller through space, such games clearly extend the literary and screen tourist experience.

Performance

Another field of activity with close synergies to mainstream literary tourism are events which allow the visitor to interact with a literary theme in ways which evoke the performative nature of tourism. Aligned to the experiential turn in tourism studies (see Chapter Three), the performance metaphor reconfigures tourists as embodied actors who perform in different settings or stages (Edensor, 2000). Light suggests that in mediatised settings such as screen or literary tourism destinations, there are powerful 'place myths' and 'this set of existing ideas about the destination will play an important role in influencing tourists' performances when they arrive there' (2009, p. 242). This interaction between the visitor and the location emphasises the dynamic nature of place which is reworked and reproduced by those who engage with it. We have already seen some examples of the performative nature of literary tourism in young visitors to Verona and fans of urban detective fiction (see Chapter Five) and in the following examples we will explore literary tourism performances which evoke the traditional sense of the term, and which offer embodied experiences where the actor/tourist becomes part of the literary landscape.

The dressing up box

The first of these examples of performance is in the use of costumes in order to take on the personae of author or literary characters. The concept of dressing up to enhance the delivery of the literary tourism experience is, of course, not a new one and guides in historical costume have been a feature of museum interpretation and battle re-enactments for generations. There are many examples of costumed literary guides, for example the entertainers dressed in traditional attire from the Spanish Golden Age on the 'Cervantes Train' from Madrid to the author's birthplace in Alcalá de Henares or the costumed guide who leads visitors around Orchard House in Concorde, Massachusetts. However, there has been an increasing focus on participants also assuming a literary persona and this may usefully be examined within the context of the evolution of 'cosplay' or 'costume play.' The term was first coined in the 1980s and is used to describe the activity of dressing up and acting as characters from popular culture whether that be comic characters, animations, video games, sci fi, or bands (Rahman, Wing-sun and Cheung, 2012). Different fan tribes emerge, and these are reflected in various conventions and shows such as the twice-yearly London Comic Con. The trend has its roots in American and Japanese popular culture and has become a global phenomenon through media convergence (Winge, 2006). Authors who have accrued a strong fan base over a long period of time and whose works have been subject to many screen adaptations seem to inspire cosplay more than others, mainly due to the merging of the imaginative with the screen presence. One very devoted and long-established tribe are the Janeites who demonstrate their devotion to Austen by dressing up in Regency costume and undertaking a wide range of related activities. Events such as the annual Jane Austen Festival in Bath with its costumed parades and balls allows for a spectacular performance of their literary fandom. Even in the more restrained world of the National Trust, who care for her home in Chawton, visitors can dress up in bonnets and gowns and attempt to write with quill and ink (see Fig. 3.1). On Canada's Prince Edward Island, erstwhile home of L.M Montgomery and her much-loved character Anne Sullivan, the most purchased souvenir is the 'Anne hat' with red braids attached which allow visitors to be transformed into the young heroine with her much-maligned ruddy hair. This performance of Anne allows visitors to become immersed in the imaginative landscape of the novels (Gothie, 2016).

Performance and events

Actual performances in which literary tourists are encouraged to spectate and even take part are devised and promoted, for example the Fiddler of Dooney Competition in Sligo, in the Northwest of Ireland. *The Fiddler of Dooney* is a poem by William Butler Yeats, first published in 1892, which celebrates the joy of the fiddler and his ability to make people dance 'like a wave of the sea.' The poem inspired the fiddle competition which started in 1965 and continued intermittently from the 1960s until 2006, from which year it has been a regular annual event. Sligo is already closely associated with the poet's life, and a statue of Yeats and family burial sites can be found there. The competition augments this with the embodiment of the fiddler in a celebration of music and literary heritage. Another literary-inspired competition, which perhaps calls for less talent, is the annual Stella Shouting Contest which has featured as part of the annual Tennessee Williams Festival in New Orleans since 2016. Participants act out the famous scene between Stella and Stanley Kowalski in Williams' play *A Streetcar Named Desire* (1947) and given the themes of the play, the organisers are now donating income generated from the event to domestic violence charities.

Perhaps one of the best-known literary events is Bloomsday, which is celebrated each year on 16 June, the day that was immortalised in James Joyce's novel *Ulysses*, published in 1922. In the text, the central character Leopold Bloom and companions travel through Dublin from 8am until the following morning. Bloomsday participants dress up in Edwardian costume and follow the route taken in the book, re-enact favourite scenes, and experience readings from the text. Fans have marked the day from as early as 1929 although the first Irish Bloomsday took place in 1954. Since then, the day continues to be a very popular, and some would say populist, event despite the notorious difficulty of getting to grips with the text itself (Keatinge, 2014). Bloomsday centres on Dublin and is hosted by the James Joyce Centre, Dublin City Council, Dublin City of Literature, and tourism organisations. This partnership clearly indicates the combination of civic pride, heritage preservation, and literary tourism promotion which is at the heart of the event. Alongside its Dublin roots, Bloomsday is also celebrated across the world in 20 locations as diverse as Melbourne, Istanbul, Los Angeles, Trieste, and Oslo (Bloomsday, 2023). One of the key elements of Bloomsday is the consumption of food and drink described in the novel. The most memorable of these meals is found at the opening of *Ulysses* where Leopold's taste for offal is established. Consequently, a breakfast of grilled mutton kidneys is an integral part of the Bloomsday experience. Fictional foods can play a key role in the performance of the literary tourism experience as can be seen in Box 9.2.

Box 9.2 Case study: Consuming fiction – the literary meal

Creating and partaking of food inspired by fiction is another means by which literature can be (literally) consumed. The fictional world is full of memorable meals such as the clam and cod chowder in Melville's *Moby Dick*, the boeuf en daube in Woolf's *To The Lighthouse*, Charles Arrowby's solitary collations in Murdoch's *The Sea, The Sea*, and the hearty picnics consumed in many of Enid Blyton's children's stories. Food provides pivotal turning points in literary plots such as the roles played by Dante's apple, Dickens' bowl of gruel, and Proust's madeleine. Recreating such fictional meals extends and embodies the literary experience and several books have attempted do this such as Dinah Fried's (2014) *Fictitious Dishes: An Album of Literature's Most Memorable Meals* or Jennifer Barclay's (2014) *A Literary Feast: Recipes Inspired by Novels, Poems and Plays.*

Commercial dining experiences based on works of fiction have become popular, for example elaborate afternoon teas are now available in London hotels, themed around the Mad Hatter's Tea Party in *Alice in Wonderland* (at the Sanderson Hotel), or Willy Wonka's confections in the *Charlie and the Chocolate Factory* Tea (at One Aldwych). Both texts clearly lend themselves well to translation into the culinary arts. The Swan restaurant beside Shakespeare's Globe Theatre in London currently serves teas inspired by *A Midsummer Night's Dream* and plans to offer one on a *Macbeth* theme. Whilst it is possible to imagine the former, with whimsical floral flavours, the latter repast is trickier to conjure up, although hopefully it will be possible to take your tea with the milk of human kindness.

Perhaps one of the most widely partaken literary meals is the Burns Supper which takes place across the world on 25 January, on the Scots poet Robert Burns' birthday. Haggis, neeps, and tatties are eaten, toasts are made, and the haggis is celebrated with a recitation of the poet's *Address to a Haggis* (1786). During the evening the songs and poetry of Burns are appreciated in a convivial atmosphere. This commemorative supper traces its history back to 1801, the fifth anniversary of the poet's death when nine of his close friends gathered in the cottage where he was born to celebrate his life (McGinn, 2011). Today an estimated nine and a half million people globally take part in a Burns Supper every year (Paterson, 2021) and these events are seen as an important means for expatriate Scots communities to assert their cultural identity.

In central Portugal, Coimbra City Council, in partnership with the University of Coimbra, launched their *Sabores da Escrita* (Flavours of Writing) initiative in 2014 which hosts literary-themed dinners, preceded by lectures. These monthly events are open to the public and explore the history and culture of food as expressed through literary works (Sabores da Escrita, 2023). In describing a literary meal, Silva suggests that 'they foster proximity between participants and literary texts, in a dialogue that stimulates the sense of appropriation of literature' (2022, 67). The sights, smells, and tastes of a literary-inspired meal provide another way to embody the writer and the text.

Literary souvenirs

In this chapter we have been exploring a number of synergistic activities which extend the literary tourism experience, whether that be watching film adaptations, visiting screen locations, or attending literary-themed events. A final area of symbiosis that will be analysed here is in the field of souvenirs which visitors take home to remind themselves of their travels. This topic has been somewhat neglected within wider tourism studies (Swanson and Dallon, 2012; Masset and Decrop, 2021) and yet it is a major touristic activity that generates considerable income.

The word 'souvenir' is originally a French verb which means to remember, and the English usage denotes an item that helps one to recall a place or event. Souvenirs may be deliberately purchased for this purpose and there is an entire industry devoted to the production of miniature landmarks, fridge magnets, and mugs which are sold in dedicated shops in tourist districts. However, souvenirs may also simply be something that a traveller has collected on their journey such as a bus ticket, a beer mat, or even a pebble or leaf.

Museums and historic sites (including, of course, literary attractions) fully entered the souvenir business in the late 1990s as both a response to reduced public funding and the increased role of consumption in leisure practices (Zitzlsperger, 2022). Fully stocked gift shops and online retail opportunities led to critiques of commodification as cultural institutions sought to exploit their collections within their souvenir ranges. Of course, as we saw in Chapter Four, taking mementoes from a literary location is by no means a recent activity and indeed the provision of commercial trinkets was considered to be necessary to protect the literary site from the sort of pilfering evidenced at Juliet's tomb in Verona, Shakespeare's chair in Stratford, and Robert Burns' hawthorn tree in Coilsfield, Ayrshire (Mackay and Pittock, 2012, Watson, 2020).

Today, literary sites and destinations offer a wide range of souvenirs, including of course the fictional works themselves in various formats, plus other items for household use, consumption, or personal adornment which will recall the visit or recreate an imaginary scene. Hemme suggests that such 'reification' strategies allow an author or character to become 'something graspable, touchable, even edible' (2005, p. 75) and she cites examples of businesses on the German Fairy Tale route selling Rumpelstiltskin cookies and baked Pied Piper rats. In her work on writer's houses, Nicola Watson suggests that the souvenir is like the literary house in miniature – 'a sort of portable museum' which conflates the real with the imaginary, and merges the reader with the writer (2020, p. 227). Thus, the motivation for taking something away from a literary tourism trip, whether it be a leaf or a T-shirt, is to embody the experience and to tuck it away for future recollection.

Although such mementoes help us to remember visits to specific locations and may be purchased with a view to supporting the site, literary-inspired gifts are also available from online retailers who recognise the attractiveness of the theme and the cultural capital that such items convey. Businesses such as The Literary Gift Company, The Literary Emporium, or Bookishly offer candles which conjure up the scents of Pemberley, Fair Verona, or Wonderland. Soap, book bags, mugs, and T shirts adorned with quotations are also widely available. Vintage-style travel posters to fictional locations such as Mordor, Neverland, and Narnia particularly underpin the ability of fiction to transport us both figuratively and literally.

Fictional characters, particularly those aimed at children, have also been the focus of 'brand extensions' where toys, clothing, and games are produced using the literary character as inspiration. A prominent example of this is Beatrix Potter's Peter Rabbit which although first created in 1902, is still inspiring soft toys, games, home furnishings, and figurines today. Potter herself understood the opportunities to make more money from her books and hand-sewed the prototype of the blue-jacketed Peter Rabbit cuddly toy which was then mass produced by the German company Steiff (Booth and Hayes, 2005). A.A. Milne's Winnie the Pooh and the Harry Potter series also continue to inspire a huge range of commercial items which extend the reach of the texts and their characters.

Contents tourism

In this chapter we have reviewed several activities which have clear synergistic relationships with literary tourism, namely the experience of screen tourism, fiction-inspired performances and events, and the purchasing of literary souvenirs. These demonstrate that the act of paying homage to an author or fictional creation is multi-layered, embodied, and can take place in settings far removed from the original literary location. We will conclude this discussion by placing the wide variety of experiences outlined above within the context of *contents tourism*. Contents tourism refers to tourist trips that have been inspired by a range of popular culture themes and the practice has been defined as 'travel behaviour motivated fully or partially by narratives, characters, locations, and other creative elements of popular culture forms including film, television dramas, manga, anime, novels and computer games' (Seaton *et al.*, 2017, p. 3). The concept was first developed in Japan in the early 2000s, when the government became aware of the potential of popular culture to increase both international and domestic tourism, under the banner of 'Cool Japan' (Seaton and Yamamura, 2015). The focus of the policy was to encourage regional development through using narrative as a resource to encourage tourism. This narrative quality should allow visitors to experience the contents or theme of a region in a multi-sensory manner (Yamamura, 2020). The contents tourism concept was therefore first developed and researched in Japan but there is now a transnational focus, evidenced by Beeton, Yamamura, and Seaton (2013) and latterly Yamamura and Seaton (2020), pioneering contents tourism scholars, whose work includes case studies from, *inter alia,* Australia, Hong Kong, Taiwan, Britain, and Switzerland.

Contents tourism (*kontentsu tsurizumu* in Japanese) attends to travel behaviours motivated by characters and narratives from popular culture, but which don't necessarily align themselves with specific media such as literary tourism or film tourism. For example, the popular Japanese art forms of manga and anime employ parody and multiple use of the same materials in different formats and consequently fans go in search of the contents rather than the specific platform or media. Consequently, on the surface, contents tourism may resemble other forms of media-based tourism, but in contents tourism studies the focus is not on the specific media but on the narrative or content itself, thus allowing an examination of the cross-media experiences which characterise contemporary tourism practice (Graburn and Yamamura, 2020). Tourist sites are considered to be media and in visiting them, tourists are 'trying not just to visit and see places themselves, but to consume, embody, reinterpret and reconstruct their favourite worlds through on-site experiences' (Yamamura, 2020, p. 10), using a media-mix across a number of different platforms.

Literary fiction and contents tourism

In considering how a contents tourism approach might help us to understand the literary tourism synergies already discussed in this chapter, it is important to consider the role of the written word which is at the heart of much of the content produced. Different media emerge over time, but this ageless medium provides a stable basis for the production of narrative worlds. Although contents tourism, like literary tourism, is concerned with narrative and character, it differs in that in literary tourism the author and the text are the main focus, whereas contents tourism 'considers the entire narrative world that combines the author's life and novels, plus all the derivative productions, related locations, and adaptations' (Sabre, 2017, p. 2). Some examples of a contents focussed study of literary tourism include an analysis of Marcel Proust tourism in France (Sabre, 2017), Jane Austen tourism in the United Kingdom (UK) (Seaton, 2020), *Heidi* tourism in Switzerland (Yamamura, 2020), and bush poet Banjo Paterson induced tourism in Australia (Beeton, 2020). In considering the many adaptations, experiences, and media now associated with Jane Austen, Seaton (2020) proposes that early visits inspired by the author were true examples of literary tourism but that there was a watershed moment in 1995, when the popular BBC television adaptation of *Pride and Prejudice,* with its iconic scene of a wet-shirted Mr Darcy, brought her work to a wider audience. After this watershed, visits to Austen's world have been inspired by so many derivative works and media that the phenomenon is better described as contents tourism. These derivative works include film, TV, and theatrical adaptations, events and festivals, fan sites, fan fiction, and numerous items of merchandise from mugs to sweatshirts emblazoned with Austen quotes. These sit alongside the biographical and imaginative sites already established on the literary tourism trail and represent the 'contentization' which expands her literary world (Seaton, 2020).

Another globally renowned literary figure, this time a fictional character, who has accrued a world of content around him is, of course, J.K. Rowling's Harry Potter and a fan of the book series may have watched the films, visited the Wizarding World theme park in Orlando, the Making of Harry Potter studio tour in Hertfordshire, or even the café in Edinburgh claiming to be the establishment where Rowling wrote her early books. They may have bought their own wand, consumed chocolate frogs, played the recently launched *Hogwarts Legacy* computer game, and even knitted their own Hedwig owl from several Potter-inspired craft kits available on the market. They can catch the 'Hogwarts Express' (otherwise known as the Jacobite Express) from Fort William to Mallaig in the Scottish Highlands and all this activity can be discussed on numerous *forae* and in the Harry Potter Fan Club. Clearly, this is an example of extreme commodification of a literary theme, but it also represents the creation of a world of content from which the contemporary traveller or content 'omnivore' (Seaton, 2020, p. 31) can produce meaning. More recently, this cultural convergence is illustrated by a commercial guidebook to the Naples of Elena Ferrante's best-selling *Neapolitan Quartet* (2012–2015) which was published in 2021. This guidebook fully embraces contents tourism in its deliberate merging of the literary, screen, and multi-sensory experience of the city so that 'each transmedia product expands the story whilst also providing access to the narrative universe for potential readers/fans' (Avalle, 2021, p. 24).

Conclusion

In this chapter, we have considered the range of synergistic experiences and products which extend the reach of literary tourism, offering a layered experience which converges media, conflates author and character, and offers opportunities for an embodiment of the literary theme. Even the most bookish literary pilgrim will find it hard to remain untouched by film and TV adaptations of favourite works and the more generalist literary tourist may actively seek out screen locations, events, computer games, and even dining experiences that allow them to inhabit the fictional world on a variety of multi-sensory levels. These offerings expand the experience offered by biographical or imaginative sites as many are not place-specific but allow the literary traveller to inhabit a wide range of imaginative worlds, assuming personae that embody the experience. Recent research on contents tourism illuminates this practice, placing the emphasis on the quality of the narrative conveyed rather than the specific medium employed. Therefore, although (literary) tourism will always be geographically determined as visitors are moved by numerous motivations to visit sites of significance to them, contents tourism reminds us that the contemporary tourist has a much wider range of experiences and modes of consumption available to them to complement and extend their practice. In the next and final chapter, we will review the journey already taken in this book and look towards the future for literary fiction tourism.

References

Agarwal, S. and Shaw, G. (2017) *Heritage, screen, and literary tourism.* Bristol: Channel View Publications.

Avalle, A. (2021) 'The ultimate Ferrante experience: Convergence culture, literary tourism, and the quest for authenticity', *Italian Culture*, 39 (1), pp. 19–32. Available at: https://doi.org/10.1080/01614622.2021.1938841 (Accessed: 12 July 2023).

Barclay, J. (2014) *A literary feast: Recipes inspired by novels, poems and plays.* Chichester: Summersdale.

Beeton, S. (2005) *Film-induced tourism.* Bristol: Channel View Publications.

Beeton, S. (2020) 'From Banjo to Basho: Poets, contents and tourism', in Yamamura, T. and Seaton, P. (eds.) *Contents tourism and pop culture fandom: Transnational tourist experiences.* Bristol: Channel View, pp. 205–223.

Beeton, S., Yamamura, T. and Seaton, P. (2013) 'The mediatisation of culture: Japanese contents tourism and pop culture', in Lester, J. and Scarles, C. (eds.) *Mediating the tourist experience.* Abingdon: Routledge, pp. 139–154.

Bloomsday (2023) Global Bloomsday. Available at: http://www.bloomsdayfestival.ie/global-bloomsday (Accessed: 3 June 2023).

Booth, E. and Hayes, D. (2006) 'Authoring the brand: Literary licensing', *Young Consumers*, 7 (1), pp. 43–53. Available at: https://doi.org/10.1108/17473610610701385 (Accessed: 1 June 2023).

Busby, G. and O'Neill, K. (2006) 'Cephallonia and Captain Corelli's Mandolin: The influence of literature and film on British visitors', *Acta Turistica*, 18 (1), pp. 30–51. Available at: http://www.efzg.hr (Accessed: 17 October 2022).

Edensor, T. (2000) 'Staging tourism: Tourists as performers', *Annals of Tourism Research*, 27 (2), pp. 322–344. Available at: https://doi.org/10.1016/S0160-7383(99)00082–00081 (Accessed: 18 September 2022).

Fried, D. (2014) *Fictitious dishes: An album of literature's most memorable meals.* New York: Harper-Collins.

The Gamer (2023) The best Agatha Christie games. Available at: https://www.thegamer.com/best-agatha-christie-games/#4-50-from-paddington (Accessed: 30 April 2023).

Gómez-Morales, B., Nieto-Ferrando, J. and Sánchez-Castillo, S. (2022) '(Re) Visiting Game of Thrones: Film-induced tourism and television fiction', *Journal of Travel & Tourism Marketing*, 39 (1), pp. 73–86. Available at: https://doi.org/10.1080/10548408.2022.2044971 (Accessed: 6 June 2023).

Gothie, S.C. (2016) 'Playing "Anne": Red braids, green gables, and literary tourists on Prince Edward Island', *Tourist Studies*, 16 (4), pp. 405–421. Available at: https://doi.org/10.1177/1468797615618092 (Accessed: 5 December 2022).

Gould, M. and Mitchell, R.N. (2010) 'Understanding the literary theme park: Dickens World as adaptation', *Neo-Victorian Studies*, 3 (2), pp. 145–171. Available at: https://neovictorianstudies.com/article/view/242 (Accessed: 19 May 2021).

Graburn, N. and Yamamura, T. (2020) 'Contents tourism: Background, context, and future', *Journal of Tourism and Cultural Change*, 18 (1), pp. 1–11. Available at: https://doi.org/10.1080/14766825.2020.1707460 (Accessed: 28 May 2023).

Hemme, D. (2005) 'Landscape, fairies and identity: Experience on the backstage of the fairy tale route', *Journal of Tourism and Cultural Change*, 3 (2), pp. 71–87. Available at: https://doi.org/10.1080/09669580508668488 (Accessed: 11 October 2022).

Jacobs, J. (2023) 'Hogwarts Legacy sells 12 million copies during dispute over Rowling', *The New York Times*, 23 February. Available at: https://www.nytimes.com/2023/02/23/arts/hogwarts-legacy-sales-rowling.html (Accessed: 18 June 2023).

Keatinge, B. (2014) '"In Dublin's Fair City": Joyce, Bloomsday, Dubliners and the invention of tradition', in Krivokapić-Knežević, M. and Nikčević-Batrićević, A. (eds.) *The beauty of convention: Essays in literature and culture.* Newcastle-Upon-Tyne: Cambridge Scholars Publishing, pp. 51–63.

Light, D. (2009) 'Performing Transylvania: Tourism, fantasy and play in a liminal place', *Tourist Studies*, 9 (3), pp. 240–258. Available at: https://doi.org/10.1177/1468797610382707 (Accessed: 29 June 2023).

Mackay, P. and Pittock, M. (2012) 'Highland Mary: Objects and memories', *Romanticism*, 18 (2), pp. 191–203. Available at: https://doi.org/10.3366/rom.2012.0084 (Accessed: 22 July 2023).

MacLeod, N., Shelley, J. and Morrison, A.M. (2018) 'The touring reader: Understanding the bibliophile's experience of literary tourism', *Tourism Management*, 67, pp. 388–398. Available at: https://doi.org/10.1016/j.tourman.2018.02.006 (Accessed: 12 December 2021).

Masset, J. and Decrop, A. (2021) 'Meanings of tourist souvenirs: From the holiday experience to everyday life', *Journal of Travel Research*, 60 (4), pp. 718–734. Available at: https://doi.org/10.1177/0047287520915284 (Accessed: 2 July 2023).

McGinn, C. (2011) 'Vehement celebrations: The global celebration of the Burns Supper since 1801', in Pittock, M. (ed.) *Robert Burns in global culture.* Lanham, ML: Bucknell University Press, pp. 189–203.

Morris, S. (2020) '"No sex, no violence, just a laugh": Mixed emotions in Cornish village as Doc Martin ends', *The Guardian*, 14 September. Available at: https://www.

theguardian.com/uk-news/2020/sep/14/no-sex-no-violence-just-a-laugh-mixed-emotions-in-cornish-village-as-doc-martin-ends (Accessed: 15 May 2023).

O'Connor, N. and Kim, S. (2014) 'Pictures and prose: Exploring the impact of literary and film tourism', *Journal of Tourism and Cultural Change*, 12 (1), pp. 1–17. Available at: https://doi.org/10.1080/14766825.2013.862253 (Accessed: 3 May 2023).

Paterson, K. (2021) 'Thousands of global Burns night suppers revealed on interactive map', *The National*, 18 January. Available at: https://www.thenational.scot/news/19018242.map-reveals-global-reach-burns-night/ (Accessed: 1 July 2023).

Rahman, O., Wing-Sun, L. and Cheung, B.H.M. (2012) '"Cosplay": Imaginative self and performing identity', *Fashion Theory*, 16 (3), pp. 317–341. Available at: https://doi.org/10.2752/175174112X13340749707204 (Accessed: 21 July 2023).

Roesch, S. (2009) *The experiences of film location tourists*. Bristol: Channel View Publications.

Sabores da Escrita (2023) Flavours of writing. Available at: https://saboresdaescrita.wixsite.com/saboresdaescrita (Accessed: 19 March 2023).

Sabre, C. (2017) 'Between literary tourism and contents tourism: Marcel Proust and his madeleine as a tourist attraction', *International Journal of Contents Tourism*, 2, pp. 1–11. Available at: https://contents-tourism.press/wp-content/uploads/2017/12/IJCT-Vol-2-Sabre-2017.pdf (Accessed: 21 August 2022).

Screen Scotland (2021) Economic value of the screen sector in Scotland in 2019. Report produced by Saffery-Champness Chartered Accountants. Available at: https://www.screen.scot/binaries/content/assets/screen-scot/funding–support/research/final_economic_value_of_screen_sector_in_scotland_report_23.6.22.pdf (Accessed: 3 June 2023).

Seaton, P. (2020) 'The contents tourism of Jane Austen's American fans', in Yamamura, T. and Seaton, P. (eds.) *Contents tourism and pop culture fandom: Transnational tourist experiences*. Bristol: Channel View, pp. 19–33.

Seaton, P. and Yamamura, T. (2015) 'Japanese popular culture and contents tourism–Introduction', *Japan Forum* 27 (1), pp. 1–11. Available at: https://doi.org/10.1080/09555803.2014.962564 (Accessed: 29 May 2023).

Seaton, P., Yamamura, T., Sugawa-Shimada, A. and Jang, K. (2017) *Contents tourism in Japan: Pilgrimages to 'sacred sites' of popular culture*. New York: Cambrian Press.

Silva, A.C. (2022) 'Literary meal', in Quinteiro, S. and Marques, M.J. (eds.) *Working definitions in literature and tourism: A research guide*. Faro: University of Algarve, pp. 67–68.

Swanson, K.K. and Timothy, D.J. (2012) 'Souvenirs: Icons of meaning, commercialization and commoditization', *Tourism Management*, 33 (3), pp. 489–499. Available at: https://doi.org/10.1016/j.tourman.2011.10.007 (Accessed: 4 May 2023).

Tourism Northern Ireland (2021) Northern Ireland Game of Thrones territory: The making of a screen tourism destination. Available at: https://www.tourismni.com/globalassets/business-development/support-by-sector/activities–attractions/grow-your-business/game-of-thrones/the-making-of-a-screen-tourism-destination.pdf (Accessed: 28 April 2023).

VisitScotland (2022a) Set in Scotland. Available at: https://www.visitscotland.org/news/2022/new-film-guide#Guide (Accessed: 26 April 2023).

VisitScotland (2022b) The Outlander Effect and tourism. Available at: https://www.visitscotland.org/binaries/content/assets/dot-org/pdf/research-insights/the-outlander-effect-tourism.pdf (Accessed: 26 April 2023).

Watson, N. (2020) *The author's effects: On writer's house museums.* Oxford: Oxford University Press.

Winge, T. (2006) 'Costuming the imagination: Origins of anime and manga cosplay', *Mechademia*, 1 (1), pp. 65–76. Available at: https://doi.org/10.1353/mec.0.0084 (Accessed: 2 May 2023).

Yamamura, T. (2020) 'Introduction: Contents tourism beyond anime tourism', in Yamamura, T. and Seaton, P. (eds.) *Contents tourism and pop culture fandom: Transnational tourist experiences.* Bristol: Channel View, pp. 1–16.

Yamamura, T. and Seaton, P. (eds.) (2020) *Contents tourism and pop culture fandom: Transnational tourist experiences.* Bristol: Channel View.

Zitzlsperger, U. (2022) 'Commerce, culture, and heritage: Souvenirs as communicators of history in the museum shop', *International Journal of the Inclusive Museum*, 15 (1), pp. 85–98. doi:10.18848/1835–2014/CGP/v15i01/85–98.

10 The future of literary fiction tourism

Over the course of this book on Literary Fiction Tourism, we have explored the relationship between the acts of reading and writing fiction and a sense of place, tracing the many different types of literary locations which entice readers to put down their books and become travellers. The development of these settings and their availability for public view constitute the 'commodification of the imagined and imaginers' (Robinson, 2002, p. 7) and in the pages of this book we have visited many places associated with 'imagined' literary locations and with the 'imaginer,' in the shape of authors' homes, haunts, and final resting places. However, a literary association can travel further than these two types of location and in a world where leisure environments are increasingly themed, fiction can also provide a unifying motif in socially-constructed literary visitor sites which may have no authentic connection to authorial biography or imaginative setting but where the power of the narrative itself provides the draw. Literary fiction tourism is not limited to individual sites and locations, however, and in these chapters we have also sought out towns and cities of literary reputation where a range of associations, including the industries supporting the production and consumption of literature, are a lure for visitors. Indeed, entire landscapes can be realised by the creation of trails and tours which overlay existing topography allowing the traveller to utilise a literary gaze. These experiences may be consumed by readers who will further augment their experiences by returning to the text or by visitors who may be inspired to seek out the novel for the first time, having had a geographical encounter with its themes. In either case, the symbiotic relationship between the text, the setting, and the reader are evident and have been used in an increasingly global context by both individual travellers and the tourism industry.

Throughout the preceding account of literary fiction tourism, this activity has been placed in its historical and contemporary context and viewed as a corollary of both cultural and heritage tourism with their attendant theoretical perspectives. One unifying theme throughout the book has been the impact of technology on the evolution of literary tourism, whether that be the technologies of transportation, publishing, or the digital world which has transformed how we read and how we travel. In this final chapter, we will

DOI: 10.4324/9781003041740-10

consider how this theme may alter the future of literary tourism, in particular the production and consumption of literature, and of literary tourism sites and their interpretation. In a post-pandemic world, we will also consider whether physical travel is even required for literary fiction to transport us to other lands.

Futures research is an activity that uses scenario planning to envisage a range of possible futures to inform strategic planning. Rather than simply forecasting the future, scenario planning 'embraces several outcomes and then, further, it employs foresight and back-casting to map out complex, layered causal powers involved in social processes and outcomes' (Robertson and Yeoman, 2014, p. 325). In their paper on the future of literary festivals, Robertson and Yeoman identify two contexts underpinning their research, namely technology and its interaction with literature and co-creation. They also highlight a number of signposts to the future including literature's role in social capital, changing roles for cultural professionals and institutions, and the characteristics of the future reader. Science fiction is also considered a useful tool for prediction and in a recent publication entitled *Science Fiction, disruption and tourism* (2021) the authors argue that, as many science fiction stories have ultimately come true, this literary genre can usefully be examined to forecast tourism's future. Science fiction will inevitably influence designers and engineers 'to turn ideas into reality' (Yeoman, McMahon-Beattie and Sigala, 2021, p. 3), for example the numerous gadgets used by the crew of the Starship Enterprise in *Star Trek* that are now in everyday usage. Science fiction has also inspired numerous visitor attractions and theme park rides, and these have been reflected in popular culture in films such as *Westworld* (1973) and *Jurassic Park* (1993) (Yeoman *et al.*, 2021) and will continue to do so. Even the moral code found in the fantasy world of Harry Potter has consequences for the future as Reichenberger (2021, p. 91) argues. With its 'ingrained messages of personal integrity, moral actions, and collective as well as individual responsibility and their subsequent positive impacts,' the Potter universe has a role to play in the unfolding transformation of tourism. In the following discussion, we will now consider the trends which may impact on the future production and consumption of literature and associated literary fiction tourism.

The future for reading and writing literary fiction

Technology has already radically changed the way in which we access and read literature and is also beginning to influence the production of fiction. In an earlier chapter, we considered the introduction of e-readers which allow us to carry around thousands of texts in one light hand-held device. Material can be purchased online and delivered to our devices within seconds. Digital Print on Demand (POD) services allow publishers to print small runs of niche titles to-order and this enables them to keep large back catalogues of texts that can be easily accessed by readers – something that would have been impossible in the era of type-setting where a print run had to be large to justify the cost. This technology is changing our attitudes to the value of the printed word as artefact. Texts, once expensive and treated with reverence and respect, can now be accessed cheaply, enjoyed, and then discarded – as the study of contents tourism proposes, the value is in the message, not the medium. To this end, a company called Short Édition have installed over 300 short story dispensers in public spaces around the world (see Fig. 10.1). Users choose their stories by length (one, three, or five minutes of reading time) and the story appears at the press of a button. Installed in shopping centres, airports, and libraries and using content created by the local community, the business claims that the dispensers are 'a creative and innovative way to boost reading for fun, encourage creative idleness and improve waiting experience' (Short Édition, 2023). In the future, a text may not even be an item separate from our own being as research into wearable devices such as contact lenses, which could deliver texts and interpretive materials straight to our eyes, suggests (Robertson and Yeoman, 2014).

Within the production of literature, digital technology has now made self-publishing much more economical, and writers can now print and distribute their works through online retailers such as Amazon. It has also facilitated the production of fan fiction – amateur re-tellings or extensions of existing texts – which are circulated and consumed amongst fan communities rather than created for commercial purposes. These works are produced without the

Figure 10.1 Short story dispensing machine, East London. Photograph by the author.

benefit of an editorial team and so the quality of writing and printing may be questionable. However, they are at least produced by the pen or keyboard of a sentient being, something which the latest artificial intelligence (AI) developments threaten. In 2022, ChatGPT (Chat Generative Pre-Trained Transformer) was launched. This chatbot (a computer programme designed to simulate human conversation) has already caused much excitement but also alarm in its ability to create natural speech in response to prompts. There is fear that it will replace jobs in the creative industries, and at the time of writing it is at the centre of a current Hollywood script writers' strike. In schools and universities, educators are rightly concerned about its use in plagiarism and several businesses have banned their employees from using the programme. The question as to whether ChatGPT can produce works of fiction is fraught – it can be prompted to write in the style of a certain author, but currently the endings and literary style that it creates are weak. The chatbot does, however, understand the mechanics of plot and AI researchers claim that the goal is for the programme to be used in tandem with humans and not as a replacement (Carter, 2023).

There are many that take a pessimistic view of the future of authorship in the light of these new developments, but they should perhaps consider that new technologies have not always replaced their antecedents – video did not, in fact, kill the radio star, nor did e-readers replace books. Even if AI programmes become skilful in creating literary texts, fascination with the human author, their life, and the magic of their craft will undoubtedly continue. It is also possible that, in the future, tired of condensed and impoverished language on social media, we may enter a post-digital era where extended literary works have even more value than they have today.

The future for literary tourism sites and landscapes

Literary tourism sites, attractions, and events continue to be developed and interest in places associated with authors and their works abides. However, new interpretation technologies are changing the way that sites are presented and how visitors interact with them. The application of augmented reality in writers' houses and literary museums, for example, is allowing the sometimes-static presentation of an author's life and work to come alive without any adaptation to the physical environment. Augmented reality uses technology to superimpose images, text, or sounds on top of what a person is already viewing. A smartphone or tablet is used to alter what the visitor sees using an application. In a writer's house, images of the author and other key figures may be provided to bring a scene to life. In a wider setting, augmented reality may be used to show how a view has altered over time, or to people a literary location with imaginary figures. Augmented reality has been used in the museum field for some years but its potential within literary tourism has yet to be fully exploited (Fenu and Pittarello, 2018). Bingemer (2021), for example, envisages a future where virtual reality can be used to create entire trips taken within the world of a literary text, using George Simenon's detective novel *Maigret and the Dead Girl* (1954), set in Paris, as an example. Over the course of a few days, small groups of 'adventurists' could be immersed in scenes from the novel taking place on the city streets as they follow Maigret in solving the mystery of the girl's identity. Accommodation, transport, and catering could be integrated into the offering which would be developed and sold by specialist tour operators in partnership with destination management organisations, thus offering a new way for visitors to experience literary cities in the future.

Digital mapping technology also offers much potential for creating more engaging and personalised literary tourism encounters. Rather than following a route on a paper map, geovisualisation, afforded by global positioning systems (GPS) and geographic information systems (GIS), allows visitors to locate themselves within the literary landscape, see images, and read text *in situ* and to interact with the cartography itself. Projects such as Google Lit Trips use Google Earth and other integrated web-based resources to augment the experience of reading, in this case primarily for educational purposes. Reader-generated mapping encourages users to upload images of literary locations and their responses to their experiences and Quinterio (2020) suggests that, in future, scoring visitors depending on the numbers of locations visited or experiences performed would lend an element of gamification to the literary tourism experience, increasing enjoyment and interaction.

New types of literary tourism sites which have recently appeared suggest that we may see more attractions in the future which integrate different elements of the literary tourism experience and make deliberate statements about the relationships between literature and place. An explicit example is found in the Seamus Heaney HomePlace (see Box 10.1) which was purposefully created to provide an entire literary landscape replete with biographical, imaginative, and reputational aspects. The Museum of Innocence in Istanbul is another example of a site which disrupts the existing literary tourism landscape. The award-winning museum was created by author Orhan Pamuk in 2012 and is based on his novel of the same name published in 2008. Both novel and museum were produced in tandem, and the museum is the realisation of the story with items that the characters use, wear, and collect in the novel carefully presented in museum displays. This deliberate pairing of site and novel plays with the spatiality of the text and the literariness of geographical space. The website advises that there is no need to have knowledge of the novel to enjoy the museum but 'just as those who read the novel will better understand the various meanings of the museum, those who visit the museum will see many things that they did not notice while reading the novel' (Museum of Innocence, 2023). It seems clear that we will see more innovative literary sites such as these in the future.

Box 10.1 Case study: At home with Seamus Heaney

The poet Seamus Heaney's HomePlace, which opened in 2016 in Bellaghy, Ulster, Northern Ireland, was designed to honour the beloved Nobel Laureate who died in 2013. The site, which was once a heavily fortified Royal Ulster Constabulary police station, was transformed into the 2,000 m^2 site at a cost of £4.24m which was mainly borne by Mid Ulster Council (Cooke, 2016).

Heaney was the author of over 20 volumes of poetry and criticism and his works received critical acclaim and were much loved by the public who voted 'When all the others were away at Mass' (from *Clearances,* 1987) as Ireland's best-loved poem of the last 100 years (Mack, 2016).

This literary site dedicated to Heaney is a symbol of the power of poetry and its association with place, but it is also an exemplar of a location which integrates the literary tourism categories which have been examined in this book as is evidenced by its name: 'Seamus Heaney's HomePlace: Life, Literature and Inspiration.' The site combines autobiographical locations, imaginative landscapes, social construction, and a traditional museological approach. The landscapes around the site have been turned into themed trails and, as an award-winning destination which attracted over 42,500 visitors from 20 countries in its first year, the site has also contributed to the literary reputation of Northern Ireland (Mid Ulster Council, 2017).

The HomePlace building is situated in Bellaghy, near the poet's birthplace in Mossbawn, and the site of his grave in St Mary's Church. The exhibition space houses the permanent collection of artefacts and photographs provided by Heaney's family. His duffel coat, school desk, and satchel feature (his writing desk is at the Bank of Ireland Cultural and Heritage Centre in Dublin) and the poems are presented against a background of video and audio contributions from King Charles, Bill Clinton, Stephen Fry, Bono, and friends and neighbours, plus recordings of Heaney himself reciting his work. The Centre also includes a 190-seat performance space called The Helicon, which is used by visitors and locals, and a library housing book collections donated by Heaney's family. Since 2021, a digital archive bringing together radio and television broadcasts presented by, or featuring, the poet has been available.

The HomePlace is linked to the surrounding landscape which was such an important source of inspiration for the poet by an elevated viewing platform which offers views across the South Derry countryside. Visitors can also venture out into Open Ground, a series of five significant locations for the poet which can be visited as a self-guided activity or as part of a guided tour. The website encourages visitors to 'discover places that shaped Seamus Heaney's life and literature – Toomebridge, Magherafelt, Moyola River, Lough Beg, and Bellaghy Bawn – and discover the smells, sights and sounds that make his poetry so uniquely of this place' (HomePlace, 2023). Interpretation panels provide information, and an app allows visitors to hear the poet read his poems in the places that inspired them. The app also provides a range of augmented reality experiences.

As an increasingly global phenomenon, literary tourism is mirroring the continuing development of more mainstream tourism activities. However, climate change and the focus on sustainable development goals mean that more responsible travel activities will feature heavily in future and the further development of more of the world's regions for tourism must be questioned. As a form of tourism which utilises local cultural themes, an existing infrastructure, and has the potential to develop more low-impact digital approaches to interpretation, tourism based on literary fiction is clearly a sustainable option for destinations who want to develop a year-round visitor economy (Carreira, 2022). Literary themes can also help to diversify domestic tourism offerings as the literary tourist does not require international travel to pursue their interests. Thus, visitor activity inspired by literature addresses the 'three pillars' of environmental, economic, and sociocultural sustainability (Purvis, Mao and Robinson, 2019) which should future-proof this most ancient form of travel.

As the world slowly emerges from the global Covid-19 pandemic of 2020 onwards, we can reflect on how we experienced literary place at a time when travel was halted. Many literary sites used digital technology to provide virtual tours and experiences to fill the gap, such as the range of activities that replaced the Bloomsday festival in 2020 and 2021 (O'Dwyer, Young and Smolic, 2022). Creative ideas helped to keep literary place alive in our imaginations during lockdown, for example in the United Kingdom (UK), the National Trust created a series of podcasts exploring the impact of the literary places in their care on the work of associated authors which are still available to download (National Trust, 2023). The spectre of the global pandemic and our current climate emergency give us pause to consider the role of international travel in the future of literary fiction tourism and we will end by considering how the imagination and local geographies can transport us when corporeal travel is either prohibited or inadvisable. Amey (2020), writing in pandemic isolation, discusses the ways in which local contexts can be brought into a text to provide imaginative ballast, for example a path through a neighbouring wood can be used to form a dramatic and atmospheric literary setting in our imagination, that same path then being forever transformed into a literary place wherever we walk there. This process means that 'actual-world places shift from shaping the reading of the text to being shaped *by* the reading of the text' (Thurgill and Lovell, 2019, p. 18). If international travel is restricted in the future, the resilience of the fiction-informed imagination means that we can continue to transport ourselves to the places and people of our most cherished literary worlds using the resources available to us.

Conclusion

In this book, we have journeyed through a series of literary fiction-inspired places and experiences, noting along the way the pervasive power of the author, the setting, and the imagination of the reader in creating the literary landscapes that travellers have been enjoying for centuries. However, the activities of travel and of producing and consuming literature have been radically transformed in recent years through the application of new technologies. In this final chapter, we have considered recent changes in the ways we access and consume literary fiction and the threats and opportunities offered by Artificial Intelligence in the creation of literary texts. These transformations have shaped the presentation and interpretation of the literary landscape and will continue to influence the future of literary fiction tourism both in real-world settings and in the virtual domain to where we may sometimes retreat through choice or necessity. But however profound these changes may prove to be in the future, it seems certain that stories, their settings, and the people who create them will continue to be an essential element of human life, offering diversion, solace, and inspiration to those who seek them out.

References

Amey, E. (2020) 'On Mirkwood, vampires and rhododendrons: Experiencing familiar places through fiction', *Literary Geographies*, 6 (2), pp. 268–273. Available at: https://www.literarygeographies.net/index.php/LitGeogs/article/view/258 (Accessed: 2 July 2023).

Bingemer, S. (2021) 'Holidays with Inspector Maigret: Mixed reality adventures as value drivers in future tourism', in Yeoman, I., McMahon-Beattie, U. and Sigala, M. (eds.) *Science fiction, disruption and tourism*. Bristol: Channel View Publications, pp. 133–144.

Carreira, V. (2022) 'Overtourism and literature', in Quinteiro, S. and Marques, M.J. (eds.) *Working definitions in literature and tourism: A research guide*. Faro: University of Algarve, pp. 111–112.

Carter, S.L. (2023) 'Can ChatGPT write a better novel than I can?', *The Washington Post*, 13 February. Available at: https://www.washingtonpost.com/business/can-chatgpt-write-a-better-novel-than-i-can/2023/02/11/e7f3d0c4-aa0e-11ed-b2a3-edb05ee0e313_story.html (Accessed: 2 August 2023).

Cooke, R. (2016) 'At home in Heaney country', *The Observer*, 2 October. Available at: https://www.theguardian.com/books/2016/oct/02/seamus-heaney-homeplace-review-bellaghy-co-down (Accessed: 1 September 2023).

Fenu, C. and Pittarello, F. (2018) 'Svevo tour: The design and the experimentation of an augmented reality application for engaging visitors of a literary museum', *International Journal of Human-Computer Studies*, 114, pp. 20–35. Available at: https://doi.org/10.1016/j.ijhcs.2018.01.009 (Accessed: 21 July 2023).

HomePlace (2023) Seamus Heaney HomePlace: Life, literature and inspiration. Available at: https://seamusheaneyhome.com/ (Accessed: 10 August 2023).

Mack, C. (2016) 'Rhyme and reason: Seamus Heaney HomePlace pays lyrical tribute to the poet', *The Guardian*, 7 October. Available at: https://www.theguardian.com/travel/2016/oct/07/rhyme-and-reason-seamus-heaney-homeplace-pays-lyrical-tribute-to-the-poet (Accessed: 10 August 2023).

Mid Ulster Council (2017) Seamus Heaney HomePlace Progress Report. Available at: Mid Ulster Council (Accessed: 4 September 2023).

Museum of Innocence (2023) About the museum. Available at: https://www.masumiyetmuzesi.org/m%C3%BCze (Accessed: 30 August 2023).

National Trust (2023) Literary podcast playlist. Available at: https://www.nationaltrust.org.uk/discover/virtual-visit/podcasts/literary-podcasts-playlist (Accessed: 1 September 2023).

O'Dwyer, N., Young, G.W. and Smolic, A. (2022) 'XR Ulysses: Addressing the disappointment of cancelled site-specific re-enactments of Joycean literary cultural heritage on Bloomsday', *International Journal of Performance Arts and Digital Media*, 18 (1), pp. 29–47. Available at: https://doi.org/10.1080/14794713.2022.2031801 (Accessed: 4 September 2023).

Purvis, B., Mao, Y. and Robinson, D. (2019) 'Three pillars of sustainability: In search of conceptual origins', *Sustainability Science*, 14, pp. 681–695. Available at: https://doi.org/10.1007/s11625-018-0627-5 (Accessed: 1 February 2023).

Quinteiro, S. (2020) 'Link up with technology application in literary tourism', in Hassan, A. and Sharma, A. (eds.), *The Emerald handbook of ICT in tourism and hospitality.* Bingley: Emerald Publishing Limited, pp. 379–389.

Reichenberger, I. (2021) 'Harry Potter and the future of tourism' in Yeoman, I., McMahon-Beattie, U. and Sigala, M. (eds.) *Science fiction, disruption and tourism.* Bristol: Channel View Publications, pp. 84–96.

Robertson, M. and Yeoman, I. (2014) 'Signals and signposts of the future: Literary festival consumption in 2050', *Tourism Recreation Research*, 39 (3), pp. 321–342. Available at: https://doi.org/10.1080/02508281.2014.11087004 (Accessed: 21 July 2023).

Robinson, M. (2002) 'Reading between the lines: Literature and the creation of touristic spaces', *Current Writing: Text and Reception in Southern Africa*, 14 (1), pp. 1–28. Available at https://doi.org/10.1080/1013929X.2002.9678112 (Accessed: 11 February 2022).

Short Édition (2023) The short story dispenser. Available at: https://short-edition.com/en/p/short-story-dispenser (Accessed: 2 August 2023).

Thurgill, J.C. and Lovell, J. (2019) 'Expanding worlds: Place and collaboration in (and after) the "text-as-spatial-event"', *Literary Geographies*, 5 (1), pp. 16–20.

Yeoman, I., McMahon-Beattie, U. and Sigala, M. (2021) 'Science fiction: A theoretical lens and methodological approach for re-imagining tourism futures', in Yeoman, I., McMahon-Beattie, U. and Sigala, M. (eds.) *Science fiction, disruption and tourism*. Bristol: Channel View Publications, pp. 3–18.

Index